ARE (PORTLAND)
SOUTHWELL
PORTLAND, DORSET
81 Ext 3391

Underwater Acoustic Systems

Macmillan New Electronics Series
Series Editor: Paul A. Lynn

Rodney F.W. Coates, *Underwater Acoustic Systems*
Paul A. Lynn, *Radar Systems*
A.F. Murray and H.M. Reekie, *Integrated Circuit Design*
Dennis N. Pim, *Television and Teletext*
Martin S. Smith, *Introduction to Antennas*
P.M. Taylor, *Robotic Control*

Series Standing Order

If you would like to receive future titles in this series as they are published, you can make use of our standing order facility. To place a standing order please contact your bookseller or, in case of difficulty, write to us at the address below with your name and address and the name of the series. Please state with which title you wish to begin your standing order. (If you live outside the United Kingdom we may not have the rights for your area, in which case we will forward your order to the publisher concerned.)

Customer Services Department, Macmillan Distribution Ltd
Houndmills, Basingstoke, Hampshire, RG21 2XS, England.

Underwater Acoustic Systems

Rodney F.W. Coates

*Professor of Electronics
School of Information Systems
University of East Anglia
Norwich*

Macmillan New Electronics
Introductions to Advanced Topics

© Rodney F.W. Coates 1990

All rights reserved. No reproduction, copy or transmission of this publication may be made without written permission.

No paragraph of this publication may be reproduced, copied or transmitted save with written permission or in accordance with the provisions of the Copyright Act 1956 (as amended), or under the terms of any licence permitting limited copying issued by the Copyright Licensing Agency, 33–4 Alfred Place, London WC1E 7DP.

Any person who does any unauthorised act in relation to this publication may be liable to criminal prosecution and civil claims for damages.

First published 1990

Published by
MACMILLAN EDUCATION LTD
Houndmills, Basingstoke, Hampshire RG21 2XS
and London
Companies and representatives
throughout the world

Typeset by
Comind (UK), Cambridge

Printed in Hong Kong

British Library Cataloguing in Publication Data
Coates, R.F.W. (Rodney F.W.)
    Underwater acoustic systems.
    1. Acoustics. Underwater
    I. Title
    620.2'5

ISBN 0–333–42541–3
ISBN 0–333–42542–1 pbk

Contents

Series Editor's Foreword	viii
Preface	ix

1 Sound Transmission Fundamentals — 1
1.1 Introduction — 1
1.2 Sound Speed — 3
1.3 The Propagation Equation — 6
1.4 Logarithmic Power Measurement — 8
1.5 The "New" Reference Unit — 8
1.6 Sound Reflection and Refraction — 10
1.7 Pressure Reflection and Transmission Coefficients — 11
1.8 Behaviour at Normal Incidence — 12
1.9 Reflection and Transmission with Varying Incidence — 13
1.10 Reflection and Transmission: the "Fast" Bottom — 14
1.11 Reflection and Transmission: the "Slow" Bottom — 15

2 The Sonar Equations — 16
2.1 Introduction — 16
2.2 Source Intensity Calculation — 16
2.3 Source Directivity — 17
2.4 Transmission Loss — 18
2.5 Target Strength — 22
2.6 Reflection Intensity Loss Coefficient — 24
2.7 Sea-floor Loss — 25
2.8 Sea-surface Loss — 26
2.9 Noise — 28
2.10 Reverberation — 29
2.11 Calculating the Signal Excess — 30

3 Characteristics and Analysis of Sonar Waveforms — 32
3.1 Introduction — 32
3.2 Swept Frequency (Heterodyne) Spectrum Analysers — 34
3.3 Filter-bank Spectrum Analysers — 36
3.4 Fast Fourier Transform Analysers — 36
3.5 Prony Analysis — 40
3.6 Further Model-building Techniques for Spectral Estimation — 41
3.7 Four-dimensional Space–Time Waveform Analysis — 43

Contents

4	**Ray Trace Modelling of Sonar Propagation**	**52**
4.1	Introduction	52
4.2	Ray Tracing Sonar Models	53
4.3	Ray Trace Calculations	56
4.4	Some Examples of Ray Modelling	58
4.5	Modelling Transmission in the Shelf-seas	65
4.6	The Lloyd Mirror Effect	70
5	**Normal Mode Modelling of Sonar Propagation** (co-authored by P.A. Willison)	**73**
5.1	Introduction	73
5.2	A Heuristic Treatment of Normal Modes in an Acoustic Waveguide	74
5.3	Normal Mode Solution for Long Ranges	80
5.4	Normal Modes as Interfering Plane Waves	83
5.5	The Normal Mode Solution Formalised	84
5.6	Normal Mode Solution for All Ranges	87
5.7	The Horizontally Stratified Channel	88
6	**Noise and Reverberation**	**90**
6.1	Introduction	90
6.2	Deep Sea Ambient Noise Level	91
6.3	The Variability of Ambient Noise with Time	93
6.4	The Variability of Ambient Noise Level with Depth	94
6.5	The Angular Distribution of the Ambient Noise Field	95
6.6	Ship-generated Noise	100
6.7	Reverberation	104
6.8	Scattering	106
7	**Acoustic Transduction**	**112**
7.1	Introduction	112
7.2	The Basic Principles of Acoustic Transduction	113
7.3	Piezo-electric Transduction	116
7.4	The Langevin Projector	117
7.5	Ring and Tube Transducer Designs	122
7.6	Resonance Behaviour of Transducers	123
7.7	Multiple Matching Layer Transducers	127
7.8	Polar Response Measurements on Transducers	129
7.9	Admittance Measurements of Terminal Response	130
7.10	Hydrophones	132

8	**Transducer Arrays**	**136**
8.1	Introduction	136
8.2	The Linear Hydrophone Array	137
8.3	The Fourier Transform Approach to Pattern Synthesis	143
8.4	Array Beamsteering	146
8.5	Directivity Index	147
8.6	The Parametric Source	147
8.7	Synthetic Aperture Sonar	149
9	**Sonar Engineering and Applications**	**153**
9.1	Introduction	153
9.2	The Basic Echo Sounder	154
9.3	Sub-bottom Profiling	159
9.4	Fishing Sonars	160
9.5	Side-scan Terrain-mapping Sonars	162
9.6	Seismic Survey	165
9.7	Acoustic Positioning and Navigation	166
9.8	Doppler Measurements	168
10	**Acoustic Communications**	**171**
10.1	Introduction	171
10.2	The Gross Attributes of the Received Signal	172
10.3	The Channel Transfer Function	175
10.4	Combating Multipath	178
10.5	Diversity Reception	178
10.6	Equalisation	181
10.7	Communication using Parametric Transmission	183
Index		186

Series Editor's Foreword

The rapid development of electronics and its engineering applications ensure that new topics are always competing for a place in university and polytechnic courses. But it is often difficult for lecturers to find suitable books for recommendation to students, particularly when a topic is covered by a short lecture module, or as an 'option'.

Macmillan New Electronics offers introductions to advanced topics. The level is generally that of second and subsequent years of undergraduate courses in electronic and electrical engineering, computer science and physics. Some of the authors will paint with a broad brush; others will concentrate on a narrower topic, and cover it in greater detail. But in all cases the titles in the Series will provide a sound basis for further reading of the specialist literature, and an up-to-date appreciation of practical applications and likely trends.

The level, scope and approach of the Series should also appeal to practising engineers and scientists encountering an area of electronics for the first time, or needing a rapid and authoritative update.

Paul A. Lynn

Preface

This text is the result of a period of some fifteen years spent both researching and teaching, primarily at Master's degree level, aspects of underwater acoustics. Its content is aimed primarily at a professional engineering or advanced undergraduate and postgraduate student audience. It is the author's intention that the book should provide a brisk, comprehensive tutorial treatment adequately referenced so that the reader may readily delve further into its subject matter. Its first two chapters are concerned with the basics of the propagation of sound in the sea and with the preliminary assessment, via the Sonar Equations, of system performance. Often, both in these first chapters and, indeed, in the remainder of the text, the treatment of practical problems — for example, in modelling propagation behaviour — is handled empirically rather than theoretically. This is because the practising engineer, or the scientist seeking to utilise underwater acoustics in, for example, oceanographic investigations, frequently needs only a first-cut visualisation as to the scope or implication of a particular task, rather than a detailed mathematical dissection of the problem.

In contrast, in Chapter 3, some detailed consideration has been given to the need for and problems associated with waveform analysis, since this is often a most important way of gaining insight into the nature of both propagation and of system performance. Recent years have witnessed a dramatic increase in the power and availability of signal processing hardware together with a decrease in its cost. Additionally, there has been a considerable broadening of the scope of algorithmic techniques available for application to signal processing tasks, to the confusion of many who might benefit from the use of such methods.

Similarly, Chapter 5 delves into the complexities of Normal Mode modelling of sound propagation in the sea. In contrast to Ray Trace modelling (covered in Chapter 4) the Normal Mode approach is far from easy to appreciate. It is, however, of profound importance in describing propagation in shallow water, or at low frequencies, as well as in sound-channels or waveguides. Chapter 5 thus purports to present a map of the territory to assist the reader in venturing further into this intricate area of computer modelling. The author would wish to express his grateful thanks to Peter Willison, also of the School of Information Systems at the University of East Anglia, who played a major part in the writing of Chapter 5.

Chapter 6 deals, again often empirically, with the subjects of noise and reverberation in the sea. Some novel material is introduced here, in discussing the angular variability of ambient noise.

The subject of acoustic transduction is covered in Chapter 7, and is followed in Chapter 8 by a treatment of the formation of groups of transducers into arrays with preferred pattern propagation characteristics. It is unfortunately the case that the subject of acoustic transducer design is almost always but poorly treated — if treated at all — in texts on underwater acoustics. Indeed, the subject as a whole is inadequately covered in the scientific and technical literature and is almost invariably described by its proponents as being 90% black art and 10% science. This, of course, is to be regretted and is, in part at least, a consequence of a policy of need-to-know on the part of Naval Laboratories and Contractors and in part the result of a lack of commitment to fundamental research in this area during the past two or three decades. Regrettably also, the subject can be done but scant justice, in a single chapter, in a text such as this.

Chapter 9 reviews the utilisation of the various techniques discussed in the preceding chapters in the construction of a range of underwater acoustic equipments. In Chapter 10, the burgeoning new field of underwater acoustic communication, which is assuming considerable importance in scientific data gathering, communication with autonomous vehicles and sub-sea oilfield control and telemetry, is also treated in some detail.

Rodney Coates

1 Sound Transmission Fundamentals

1.1 Introduction

The science of acoustics involves the study and practical application of sound transmission in solid and fluid media. Although the subject is one of considerable scope, as figure 1.1 illustrates, our interest will lie particularly in applications involving sound transmission in the sea and in its underlying sediment layers and rock strata. Sound transmission is the single most effective means of directing energy transfer over long distances in seawater. Neither radio-wave nor optical propagation is effective for this purpose, since the former, at all but the lowest usable frequencies, attenuates rapidly in the conducting salt water and the latter is subject to scattering by suspended material in the sea. Underwater acoustics is thus a topic of extreme importance in military and commercial applications.

Sound is a longitudinal wave motion which can exist in any compressible transport medium. Imagine the transport medium to resemble a three-dimensional lattice of elastically interconnected "particles". Suppose that one particle is displaced from its rest position and then released. The elastic interconnection between the particles will allow a disturbance to propagate outwards from the location of the initial displacement.

The key factors which describe the propagation, physically, are **particle velocity** and the local **pressure** which is responsible for creating particle displacement. A mathematical study of the physics of sound leads to the formulation of "wave equations" which are differential equations interrelating particle velocity and pressure. These equations, which we shall examine in greater detail later, incorporate as a "constant of proportionality" a quantity which determines the rate at which a disturbance propagates through the medium. This quantity is the **speed of sound** and is an important characteristic of all physical media which sustain sound propagation and of all engineering materials used in equipments for the generation or detection of sound.

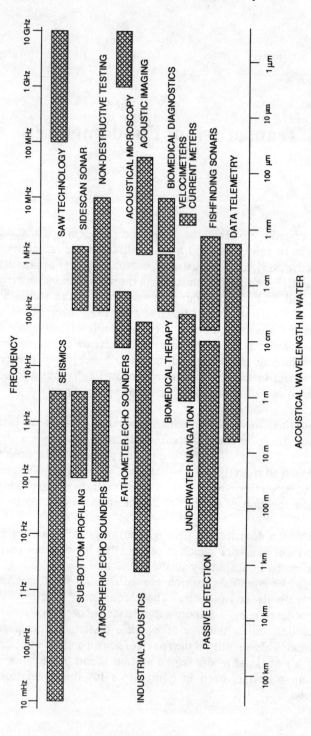

Figure 1.1 The scope of acoustical engineering

1.2 Sound Speed

Sound speed, conventionally denoted by the letter "c" is, itself, determined (through the wave equations) by three other physical properties of the medium, namely its **specific heat at constant pressure**, γ, its **density**, ρ, and its **isothermal bulk modulus of elasticity**, B. The inter-relationship between these various quantities is given by the following equation, which is attributed to Newton

$$c = (\gamma B/\rho)^{1/2}$$

In distilled water at 20°C and at standard atmospheric pressure, the physicist measures γ as 1.004, ρ as 998 kg m^{-3} and B as 2.18 x10^9 N m^{-2}. We thus calculate sound speed as 1481 m s^{-1}.

In practice, constraints in measurement accuracy of γ, B and ρ limit the value of this equation in providing a prediction of sound speed. It is, perhaps, sufficient to note that all three variables are quantities which depend upon **temperature**, T, **pressure** P and, for real sea-water, chemical composition. Chemical composition is classically expressed in terms of **salinity**, S, and more recently in terms of electrical **conductivity**, G. Consequently, it has been argued that sound speed may be expressed, to adequate accuracy, as some suitable function of temperature, pressure (or depth, Z) and salinity (or conductivity):

$$c = f(T,P,S)$$

Many polynomial approximations have been formulated to yield suitable expressions to satisfy such a relationship. These polynomials have been derived from the results of experimental measurements of sound velocity by curve fitting. The measurements are performed under carefully controlled laboratory conditions using precision "primary standard" sound velocimeters. Less accurate, but smaller and more robust "secondary standard" velocimeters are used for field measurements.

Early measurements on distilled water at atmospheric pressure performed at the US Bureau of Standards were used to produce the fourth-order polynomial

$$c = 1402.736 + T(5.03358 + T(-0.0579506 + T(3.31636E{-4} + T(-1.45262E{-6} + 3.0449E{-9}))))$$

Later work on distilled water at elevated pressures and on carefully prepared "standard" sea-water samples led to the formulation of a bewildering array of sound speed equations. At the present time, the benchmark equation should be taken as the Lovett [1.1] expression:

$$c = 1402.394 + T(K_1 + S(K_2 + K_3SP) + K_4P^2 + T((K_5 + K_6S) + K_7T)) + K_8S + P(K_9 + P(K_{10} + P(K_{11}T + K_{12}S)))$$

where

$K_1 = 5.01132$ $K_2 = -1.266383E-2$ $K_3 = 2.062107E-8$
$K_4 = -1.052396E-8$ $K_5 = -5.513036E-2$ $K_6 = 9.543664E-5$
$K_7 = 2.221008E-4$ $K_8 = 1.332947$ $K_9 = 1.605336E-2$
$K_{10} = 2.12448E-7$ $K_{11} = 2.183988E-13$ $K_{12} = -2.253828E-13$

and T is measured in degrees Centigrade, S in parts per thousand (‰ or ppt) and P in decibars. To assist in computer coding this and other key formulae in this text, set-point test values are provided. Check, for $T = 14°C$, $S = 35‰$ and $P = 100$ decibars that $c = 1505.100285$. In using the above formula, to obtain highest accuracy, it is necessary to correct the inter-relation between depth and pressure for geographical latitude θ. A useful "universal" formula for the inter-relation of pressure, P, in decibars and depth, z, in metres is given [1.2] by

$$P = 1.0052405(1 + (5.28E-3)\sin^2\theta)z + (2.36E-6)z^2$$

An alternative formula permits calculation of pressure in kg cm^{-2} rather than decibars:

$$P = 1.04 + 0.102506(1 + 0.00528 \sin^2\theta) z + (2.524E-7)z^2$$

and check that $P = 104.0690158$ for $\theta = \pi/4$ and $z = 1000$ m.

Finally, we note that recent developments in oceanographic measurement practice demand the measurement of salinity via conductivity, G, which is measured in Siemens per metre (S m^{-1}), this parameter being more amenable to direct, rapid in-situ observation. A useful but perhaps not definitive relationship is given by the "Collias" equation

$$S = -0.505 + 11.15294G + 0.3680067G^2 - 0.35412GT - 0.0120291TG^2 + 0.0086GT^2 + 0.0000048G^2T^3 - 0.00011GT^3$$

and check that $S = 33.215408‰$ for $G = 4$ S m^{-1} and $T = 14°C$ (1 S m^{-1} = 10 mmho cm^{-1}).

Practical estimation of sea-water sound speed may thus be made by measuring temperature, depth and salinity. For military purposes, which involve primarily the measurement of sound-speed versus depth profiles, and which are important in ray-trace sonar prediction programs, only temperature and depth need usually be measured. Salinity would be taken at the "typical" value of 35‰. The commonest way of acquiring this information is by means of a bathythermograph, particularly the "expendable" kind. This latter equipment, shown in figure 1.2, consists of a bomb-shaped projectile containing a thermistor connected to a surface electronics package by a spool of fine wire. The wire plays out as the bomb falls through the water column and snaps at maximum extension. The bomb rate of fall is predictable, so that depth can be inferred at any known time after launch. The resistance of the thermistor measured via the connecting wire, varies with, and thus measures sea-water temperature.

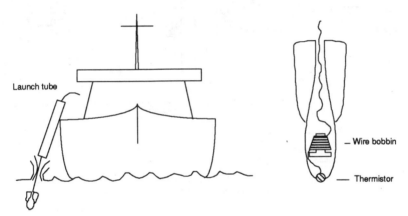

Figure 1.2 The Expendable Bathythermograph: XBT

For more accurate measurements, such as might be required in the offshore survey industry, or for definitive measurements in oceanographic acoustics, either a "CTD" or a field sound velocimeter would be employed. The term "CTD" is used to describe a nest of sensors, often employed in oceanographic surveys, which measure conductivity temperature and depth. Confusingly, the "C" in "CTD" stands not for sound speed but for conductivity for which, as we have seen, the internationally agreed abbreviation is actually G. Field measurement accuracy on conductivity is typically ±0.03 S m^{-1} (±0.3 mmho cm^{-1}), being equivalent to an accuracy of 1% of full scale on a salinity scale extending to 35‰ and thus equivalent to a full-scale conductivity of 3 S m^{-1}. A field measurement accuracy for temperature is taken to be ±0.1°C and for pressure 1% of stated depth. The accuracy of estimation of sound speed by conversion is thus depth dependent. At the surface, and at depths to a few hundred metres the accuracy on estimation is of the order of ±0.6

m s^{-1} (±400 parts per million) rising modestly to ±0.8 m s^{-1} (530 parts per million) at a depth of 4000 m. The field sound velocimeter [1.3], [1.4] illustrated in figure 1.3 operates on an acoustic pulse "sing-around" principle. A high-frequency acoustic pulse is launched into the water sample by a transmit/receive transducer. It traverses a folded 10 cm path, returning to retrigger a new pulse. The pulse repetition frequency is thus ten times the speed of sound in metres per second. The folded path is used to minimise the effect of errors which could be caused by water flow in the vicinity of the instrument. Typically, a sound velocimeter should be able to measure sound speed to an accuracy of the order of one part in 10^5. Sound velocimeters have the disadvantage of being expensive and extremely difficult to calibrate and maintain, despite their apparent simplicity.

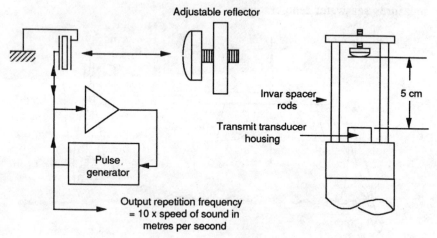

Figure 1.3 The sing-around secondary standard or "field" velocimeter

1.3 The Propagation Equation

Sound speed is the critical parameter which inter-relates transmission frequency and wavelength. Sound is a **travelling wave**. This means that, to an observer at a fixed location, sound pressure fluctuates as a function of time. However, to an observer with a global view of the pressure field, the propagating wave is also a function of spatial displacement from the sound source. For a pulsed sinusoidally-varying sound wave emanating as a plane wave-front in the horizontal, we write

$$p(t,x) = A(t + kx/\omega)\sin(\omega t + kx)$$

where A(t) is the **envelope** of the pulse shape, ω is the **radian frequency** and k is the (horizontal) **wave number**. The following equations apply:

$$\omega = 2\pi f; \quad c = f\lambda; \quad T = 1/f; \quad k = \omega/c = 2\pi f/c$$

where f is the **frequency** measured in **Hertz** (Hz) or "cycles per second". T is the **temporal period** and λ is the **wavelength**. At a fixed location, for example at x = 0, the signal has a temporal dependence

$$p(t) = A(t)\sin \omega t$$

and at a fixed instant of time, for example at t = 0, a spatial dependence

$$p(x) = A(kx/\omega) \sin kx$$

A plausible envelope for a sonar "ping" is the gaussian pulse shape

$$A(t + kx/\omega) = \exp(-0.72\,(\pi B(t + kx/\omega))^2)$$

where B is the **signal bandwidth**, determined primarily by the transducer(s) used to convert acoustic signals to electric signals, or vice versa. Typically B ≈ 0.1f. The gaussian envelope and the resulting time and spatial functions corresponding to the sonar "ping" are shown in figure 1.4.

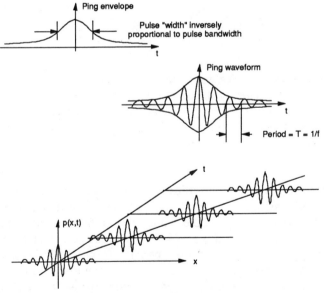

Figure 1.4 The sonar "ping" waveform

1.4 Logarithmic Power Measurement

Acoustic power levels discernible by the human ear span an extremely wide range. For example, the intensity of the sound made by the proverbial "drop of a pin" is about one millionth of a millionth (10^{-12}) of the intensity you might hear if you listen to a jet aircraft taking off. In order to draw this range within a more handleable compass, a logarithmic measure of power, known as the **decibel** is widely used. The decibel, a dimensionless unit, for which the abbreviation "dB" is used, is defined in terms of the ratio of a measured power to a reference power, thus

$$\text{dB ratio} = 10\log_{10} (\text{measured power/reference power})$$

Thus, if measured and reference powers are equal, their ratio is unity and the dB ratio is zero. If the measured power is ten times the reference power, then the power ratio is 10 and the dB ratio also 10. The table given below extends this example

power ratio:	1/100	1/10	1	10	100	1000
dB notation:	−20	−10	0	+10	+20	+30

When the decibel unit is used, the scaling of power, to account for example for loss during transmission, is achieved by addition, rather than by multiplication.

1.5 The "New" Reference Unit

Underwater acoustics has always been bedevilled by a plethora of confusing systems of units. This problem has been exacerbated by the frequent intrusion of Naval terminology so that, in any one document, horizontal range may be found measured in kiloyards whilst depth is referred to in feet, kilofeet or fathoms, speed of sound in feet per second and target speed in knots. We shall adopt SI units in this text.

Acoustic waves in the sea are pressure waves. Pressure is force per unit area. In the SI system, force is measured in Newtons, with one Newton being (about) the force exerted on the palm of your hand by an apple placed thereon. Pressure would thus be measured in Newtons per square metre and the unit of pressure is referred to as the Pascal, for which the abbreviation "Pa" is used. A pressure of 1 Pa is far larger than one would ever expect to encounter in normal underwater acoustic operations. Consequently, pressure is typically measured in units of one millionth of a Pascal, which is referred to as one microPascal, and abbreviated to 1 µPa.

Sound Transmission Fundamentals

By way of providing an illustration to give a "feel" for the practical significance of such fluctuations, a typical underwater hydrophone, sensing a sound field with pressure fluctuation of 1 µPa, could be expected to generate at its terminals a waveform with an amplitude of 100 picovolts (10^{-10} volts). Such a hydrophone would be said to exhibit a sensitivity of –200 dB relative to a standard of 1 volt per µPa.

The pressure wave discussed above, p(t,x), describes the instantaneous fluctuation of pressure as a function of time and spatial displacement. As it happens, the sonar "ping" depicted above is a waveform of a type referred to as a finite energy waveform. Waveforms which are repetitive, such as the "ping...ping...ping..." of an echo sounder, or which are continuous in nature, such as bio-acoustic emanations or sea noise, are referred to as finite power waveforms. For waveforms in this latter class, it is preferable to think not in terms of instantaneous pressure fluctuation, but in terms of an averaged pressure fluctuation. We use the **mean square** pressure, since this quantity is proportional to the time averaged power transferred by the wave. In fact the most useful measurement, derived from the average power transfer, is the power transferred per unit area normal to the direction of propagation. This quantity, measured in watts per square metre, is the **acoustic intensity**.

The acoustic intensity of any pressure wave is measured in decibels relative to the intensity of a reference wave. The reference wave is taken to be a plane wave, of root mean square pressure equal to one microPascal. The student of acoustics will frequently encounter, in the literature, the terminology "N dB re 1 µPa". This means that the intensity of the measured pressure wave is greater by N decibels than the intensity of the reference pressure wave. Notice that the terminology misses out, for brevity, an important part of the definition. It should more correctly read "N dB re (the intensity of a plane waveform of rms pressure equal to) 1 µPa". The part of the definition in the brackets should always be borne in mind.

The actual acoustic intensity of the reference wave is calculated from the formula

$$I = \overline{p^2(t)}/\rho c$$

where the product ρc is referred to as the **acoustic impedance** of the transport medium, is denoted σ and is ascribed the unit of the **Rayl**. Thus, for the plane wave of rms pressure 1 µPa, we calculate, knowing that ρ = 1000 kg m^{-3} and c = 1500 m s^{-1}, or σ = 1.5 MRayl, an intensity of 0.67E – 18 W m^{-2}.

By way of example, suppose that an acoustic source is supposed to exhibit an output intensity of +120 dB re 1 μPa. We may calculate the power ratio as

$$\text{power ratio} = 10^{(\text{dB ratio}/10)}$$

and in this case, the power ratio is $10^{(120/10)} = 10^{12}$. The actual intensity of the source is thus 0.67E–6 **watts per square metre**.

1.6 Sound Reflection and Refraction

Sound reflection, that is, **specular** ("mirror-like") reflection, obeys the same law as in geometric optics, figure 1.5, with

$$\theta_1 = \theta_3$$

Sound refraction obeys Snell's law, with

$$\sin\theta_2/\sin\theta_1 = c_2/c_1$$

Transmission will always take place from lower to higher acoustic impedance. For example, sound will always penetrate from air to water, irrespective

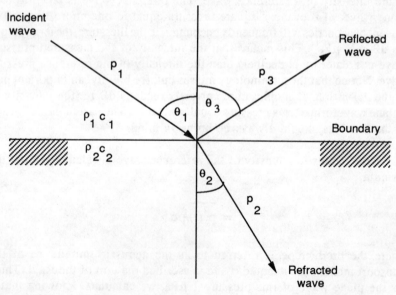

Figure 1.5 Reflection and refraction at the boundary between acoustically dissimilar media

of angle of incidence. Total internal reflection can occur (if the angle of incidence is inclined sufficiently far from the normal) if transmission is attempted from a medium of higher to one of lower acoustic impedance. The reader should recall, from geometric optics, that the critical angle of incidence, θ_c, occurs when, as the incident ray swings away from the normal, the angle θ_2 made between the emergent ray and its normal increases to graze along the interface, so that $\theta_2 = 90°$. This marks the onset of internal reflection. Then, $\sin\theta_2 = 1$ and $\theta_1 = \theta_c = \sin^{-1}(c_1/c_2)$.

1.7 Pressure Reflection and Transmission Coefficients

The acoustic impedances of the materials on either side of a boundary determine the degree of reflection or transmission across the boundary. Such properties are important in designing transducers, in determining sea-floor sediment properties acoustically, in sonar modelling and in assessing target strength. **In this section, the materials on either side of the boundary are assumed lossless.** In section 2.7 an empirical formula which is not restricted by this presumption is presented to describe, for practical modelling purposes, loss in acoustic intensity following reflection from the sea-floor. For a detailed analysis of the derivation of the following results, the reader is referred to Brekhovskikh [1.5] and to Clay and Medwin [1.6]. Inclusion of sea-floor attenuation, particularly, in predicting the pressure reflection and refraction coefficients, is due to Mackenzie [1.7].

The **pressure reflection coefficient** at a boundary is given in terms of the incident angle θ_1 by

$$R_{12} = p_3/p_1 = (A - B)/(A + B)$$

and the **pressure transmission coefficient** is given by

$$T_{12} = p_2/p_1 = 2A/(A + B)$$

where

$$A = \theta_2 \cos\theta_1 \text{ and } B = \sigma_1 \cos\theta_2$$

Notice that if c_1 and c_2 have the same value (but could none the less be properties of materials of differing density and thus differing acoustic impedance — a phenomenon observed in, for example, some sea-floor sediments) — then θ_1 and θ_2 will have the same value and both reflection and transmission coefficients will exhibit values which will be independent of angle of incidence. These values will be $R_{12} = (\rho_2 - \rho_1)/(\rho_2 + \rho_1)$ and $T_{12} = 2\rho_2/(\rho_2 + \rho_1)$.

1.8 Behaviour at Normal Incidence

If $\theta_1 = 90°$ then the reflection and transmission coefficients become, respectively

$$R_{12} = (\sigma_2 - \sigma_1)/(\sigma_2 + \sigma_1) \quad \text{and} \quad T_{12} = 2\sigma_2/(\sigma_2 + \sigma_1) = 1 - R_{12}$$

The fact that $T_{12} = 1 - R_{12}$ reflects a condition inherent in the derivation of the formulae, that the interface itself sustains no excess pressure.

Two extreme conditions are of immediate interest. For an upward-going soundwave, normally incident on the (inner) sea-surface, σ_2 becomes the acoustic impedance of air and σ_1 that of water. We note that $\sigma_2 \ll \sigma_1$, so that $R_{12} \approx -1$ and $T_{12} = 0$. We refer to the sea-surface as being a "pressure-release" boundary and note that phase-inversion of a reflected wave is to be expected.

By contrast, a sound-wave, normally incident on, say, a granite sea-wall or a thick steel plate so that $\sigma_2 \gg \sigma_1$, will exhibit $R_{12} = +1$ and $T_{12} = 0$. We note that phase inversion on reflection does not now occur but that reflection is again virtually total, with no transmission into the second medium from the water.

Next consider the case when $\sigma_2 = \sigma_1$. This condition might be considered to represent the circumstance of sound incident upon a "rho-cee" rubber membrane such as is used to protect some underwater transducers ("rho-cee" rubber is a rubber designed specifically to have the same acoustic impedance as water). For this case, $R_{12} = 0$ and $T_{12} = +1$; transmission is total and no sound is reflected.

At the sea-floor, sediment acoustic impedance ranges from about 4 MRayl, for coarse sand, to about 2 MRayl, for fine abyssal clays. The corresponding range of reflection coefficients lies between +0.45 and +0.14 and the range of transmission coefficients therefore lies between +0.55 and +0.86.

It is of some interest at this point to consider the power or energy reflected from or passing across the sediment interface; that is, we wish to determine an intensity reflection or transmission coefficient. Because the reflected wave remains within medium 1, the intensity reflection coefficient, μ_r, is simply the square of R_{12} and will be given as

$$\mu_r = I_3/I_1 = ((\sigma_2 - \sigma_1)/(\sigma_2 + \sigma_1))^2$$

and, because the sum of reflected and transmitted powers must be unity, the transmission coefficient will be

$$\mu_t = I_2/I_1 = 1 - \mu_r = 4\sigma_1\sigma_2/(\sigma_1 + \sigma_2)^2$$

It thus follows that, for $\sigma_1 = \sigma_2$, power flow is total, as we should hope. For our previous examples, we find power flow into the sediment of between 80% (for coarse sand) and 98% (for fine abyssal silt).

1.9 Reflection and Transmission with Varying Incidence

We note first that, for angles of incidence less than critical, total internal reflection will not occur and transmission and reflection will both (in general) be present. In our previous expressions, B will be real and of value

$$B = \sigma_1(1 - (c_2/c_1)^2 \sin^2 \theta_1)^{1/2}; \; \theta_1 < \theta_c$$

Both the reflection and transmission coefficients will thus be real and will vary in value with changing angle of incidence. No phase shift will occur.

For angles of incidence equal to the critical angle, $\theta_1 = 90°$ so that $\cos\theta_1 = 0$ and thus T_{12} is zero. This remains true for angles greater than critical since then total internal reflection occurs and no transmission across the interface between the two media can take place. At critical incidence, B becomes zero.

Above critical incidence, B becomes imaginary but may for convenience be re-written as

$$B' = \sigma_1((c_2/c_1)^2 \sin^2 (\theta_1) - 1)^{1/2}; \; \theta_1 > \theta_c$$

so that, in turn, the reflection coefficient may be expressed as the ratio of complex conjugates

$$R_{12} = (A - jB')/(A + jB')$$

It then follows that the magnitude of the reflectivity of the surface is invariant with angle of incidence: $|R_{12}| = 1; \; \theta_1 > \theta_c$.

The reflected wave exhibits, however, a phase lag determined by the ratio of the imaginary to the real part of the numerator or denominator of the reflection coefficient

$$\varphi = -2\tan^{-1}\{(\sin^2\theta_1 - (c_1/c_2)^2)^{1/2}/((\sigma_2/\sigma_1)\cos\theta_1)\}; \quad \theta_1 > \theta_c$$
$$= 0; \quad \theta_1 < \theta_c$$

The reflection coefficient may then be written in polar form, as

$$R_{12} = \exp(\varphi); \quad \theta_1 > \theta_c$$
$$= (A - B)/(A + B); \quad \theta_1 < \theta_c$$

If we return to the problem of reflection from the sea-surface from within, we note that $\sigma_1 \gg \sigma_2$ and that $\rho_1 \gg \rho_2$. Our expression for φ thus reduces to $\tan(\varphi/2) \Rightarrow \infty$, or $\varphi \Rightarrow 180°$. That is, irrespective of the angle of incidence, a surface-reflected wave will always suffer phase inversion.

1.10 Reflection and Transmission: the "Fast" Bottom

Many realistic sea-bed conditions are represented by the so-called "fast" bottom, where both sound speed and density significantly exceed those of the overlying water. Figure 1.6 illustrates the amplitude and phase functions for the pressure transmission and reflection coefficients for such a sea-bed.

Here, the bottom is presumed to have a high (for a sedimentary deposit, at least) sound speed of 1800 m s^{-1}. The higher density values depicted correspond loosely to the coarser sands. As can be seen, the critical angle does not change, being dependent only on the ratio of sound speeds in the two media. For angles greater than critical, total reflection occurs. For angles less than critical, sound passes into the sediment. The greatest transmission occurs at normal incidence, when θ_1 is zero. The reflected and transmitted waves exhibit zero phase shift for angles of incidence which are less than critical, and a phase shift rising to 180° at grazing incidence.

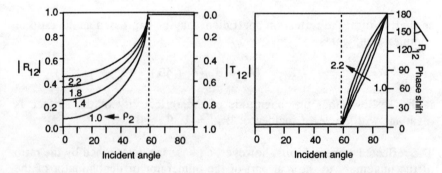

Figure 1.6 Reflection and transmission pressure coefficients versus incident angle θ_1: the "fast" bottom

1.11 Reflection and Transmission: the "Slow" Bottom

Under some circumstances, even though the sediment density will exceed the density of the overlying sea-water, the sound speed may yet be less. This is the condition referred to as a "slow" bottom. Now it is possible for "intromission" to occur. Figure 1.7 illustrates this phenomenon. We see that, at an angle

$$\theta_i = \sin^{-1}((\sigma^2_2 - \sigma^2_1)/c^2_2(\rho^2_2 - \rho^2_1))^{1/2}$$

perfect transmission into the sediment may occur, the reflection coefficient being zero.

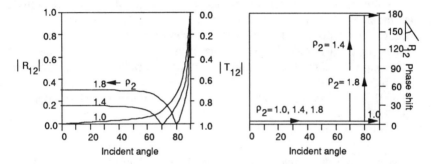

Figure 1.7 Reflection and transmission pressure coefficients versus incident angle θ_i: The "slow" bottom

References

[1.1] J.R. Lovett, Merged Sea-Water Sound Speed Equations, *J. Acoust. Soc. Am.*, Vol. 63, No. 6, 1978, pp. 1713-1718

[1.2] C.C. Leroy, Development of Simple Equations for Accurate and More Realistic Calculation of the Speed of Sound in Sea Water, *J. Acoust. Soc. Am.*, Vol. 46, No. 1, 1969, pp. 216-226

[1.3] R.L. Williamson, G. Hodges and E. Eady, A New Sound Velocity Meter, *The Radio and Electronic Engineer*, June 1967, pp. 387-393

[1.4] K.V. Mackenzie, A Decade of Experience with Velocimeters, *J. Acoust. Soc. Am.*, Vol. 50, No. 5, Pt 2, pp. 1321-1333

[1.5] L.N. Brekhovskikh, *Waves in Layered Media*, Academic Press, New York, 1960

[1.6] C.S. Clay and H. Medwin, *Acoustical Oceanography*, Wiley, New York, 1977

[1.7] K.V. Mackenzie, Bottom Reverberation for 530 and 1030-cps Sound in Deep Water, *J. Acoust. Soc. Am.*, Vol. 33, No. 11, 1961, pp.1498-1504

2 The Sonar Equations

2.1 Introduction

Underwater acoustic systems inevitably involve the detection of signals. The fundamental criterion which determines the effectiveness of all detection processes has to do with determining the extent to which the received signal exceeds, or is swamped by, such corrupting influences as may exist. In electromagnetic detection equipments – radio and radar systems – antenna-borne signals are of such small size that the significant corrupting influence may well be the similarly small, random, gaussian noise waveforms deriving from charge transport processes or molecular agitation in the antenna and front-end receiving circuits.

Whilst it is true that sonar systems also can be subject to a predominating gaussian noise corruption, the fact that the sea is bounded by excellent reflectors – its surface and floor – means that reverberation (multiple delayed echoes of the transmitted signal) may in some instances present a far greater problem. The sonar equations provide a variety of statements of sonar system efficacy. They are established by combining information pertaining to source power output and directivity, transmission loss and noise or reverberation corruption. The sonar equations produce – let it be said – guidelines, rather than exact results. The sea is a complex transmission medium and the designer, in applying the sonar equations, would do well to err in the direction of safety, in setting signal levels, and in the direction of pessimism, in predicting loss or corruption level.

2.2 Source Intensity Calculation

The **source intensity** is a measure, in dB re 1 µPa, of the power flux [W m^{-2}] delivered into the water by a source and is always referred to **standard range** from a presumed **acoustic centre** of the source. Standard range is 1 metre. The acoustic centre is a convenient fiction which provides a starting location for loss calculations. At 1 m, the acoustic centre is surrounded by a spherical envelope of area $4\pi r^2 = 12.6$ m^2. If the source power output is P watts, and the source radiates equally strongly in all directions, then the source intensity at standard range is P/12.6 W m^{-2}. The relative acoustic intensity, denoted SL and measured in dB re 1 µPa, is calculated as

$$SL = 10 \log_{10}((P/12.6)/\text{reference wave intensity})$$

Recalling the result for reference wave intensity from section 1.5, we find that

$$SL = 10 \log_{10}((P/12.6)/0.67\text{E}{-}18)$$

$$= 10 \log_{10} P + 167$$

Notice that source intensity can usually only be measured remotely from the source and in deep water, to avoid reverberation problems. Scaling of the remote measurement back to 1 m reference range must be accomplished by taking into account spreading and loss laws within the water.

2.3 Source Directivity

Our calculation of source intensity assumes omnidirectional spreading of sound from its acoustic centre. Often, like the light beam produced by the focusing mirror of an electric torch, the sound does not spread omnidirectionally. If the beam subtends ϕ steradians of solid angle, then the power concentration, for given source power output, in the beam and relative to the acoustic intensity which would have been generated, had spreading indeed been omnidirectional, is scaled by a factor

$$4\pi/\phi$$

This scaling factor is the most elementary description of a source parameter known as the **directivity index**. It is customary to express directivity index in decibels, since it represents an intensity ratio. We write

$$DI = 10 \log_{10}(4\pi/\phi)$$

The beam solid angle depends upon the dimensions of the source, relative to a wavelength, at the frequency of interest. It, and consequently directivity index, may thus be quite difficult to envisage, let alone assess for sources, such as ships, which emit broadband radiation and are of complex mechanical structure and surface geometry. We shall address the problem of estimating beam angle for transducers in chapter 7 and arrays of transducers in chapter 8. For some sonar projectors, which are usually narrowband and of well defined acoustic aperture, beam angle may be quite easily defined. For example, for the circular piston transducer of diameter D, the directivity index is calculated as

$$DI = 20 \log_{10}(\pi D/\lambda)$$

so that increasing the number of wavelengths encompassed by the diameter increases the directionality and thus improves the directivity index.

Finally, calculation of source intensity, if modified to reflect the concentrating effect of limited solid beam angle, must be amended to

$$SL = 10 \log_{10} P + 167 + DI$$

2.4 Transmission Loss

It is frequently of value to be able to assess the accumulated decrease in acoustic intensity, as a pressure wave propagates outwards from a source. The parameter which describes the decrease of intensity with distance is known as the **transmission loss**, and is denoted TL. The transmission loss is the sum of a **spreading loss** and an **attenuation**, the latter caused by the unavoidable frictional conversion of sound into heat during propagation.

The most fundamental spreading-loss law is that which describes **spherical** or **free-field** spreading. Free-field conditions will be approximated only when all reflecting boundaries are so far from the source and receiver that no channelling of acoustic energy can occur. At low frequencies this will typically be the case only in deep water. At very high frequencies, because attenuation per unit distance rises with increasing frequency the effect may also be evident in shallow water. The basic loss law for spherical spreading, the "inverse square law", giving the intensity I(R) at range R, relative to intensity at 1 m standard reference range, is

$$I(R) = R^{-2}$$

or, expressed in dB

$$I(R) = 20 \log_{10} R$$

If reflections from sea-surface and sea-floor result in the sea behaving like an acoustic waveguide, free-field conditions will not pertain. Propagation may then take place with a **cylindrical** spreading law, for which

$$I(R) = R^{-1}$$

or, in dB

$$I(R) = -10 \log_{10} R$$

Since boundary reflections vary with sea-state and bottom material properties, and since sound penetration into the sea-bed at low frequencies is quite good, the cylindrical law tends to under-estimate loss. A "practical" law, intermediate between the spherical and cylindrical laws, is thus often invoked for "first-cut" calculations in sonar system design. The practical law is defined by the expression

$$I(R) = -15 \log_{10} R$$

Attenuation in the sea is caused mainly by **viscous friction** and at frequencies in excess of (about) 1 MHz, loss is as measured for distilled water [2.1].

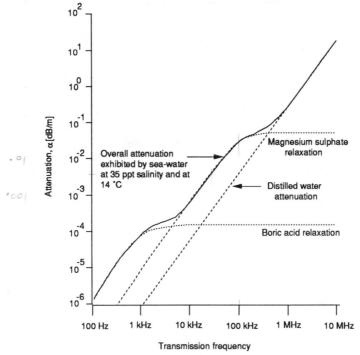

Figure 2.1 Attenuation in distilled water and in sea-water, showing the increase in attenuation brought about by molecular resonance effects

As frequency is decreased, **molecular resonance effects** conspire to worsen the attenuation figure for sea-water, by comparison with the distilled water value. At frequencies below (about) 500 kHz, the presence of magnesium sulphate, in solution in sea-water, begins to intrude an excess attenuation over the distilled water loss, ultimately increasing the attenuation uniformly, by a factor of about eighteen for frequencies below (about) 70 kHz [2.2]. At frequencies below (about) 700 Hz, boric acid – despite its small concentration in sea-water – adds in a further uniform sixteeen-fold increase in loss [2.3].

The graphs presented in figure 2.1 illustrate the variability of sea-water attenuation factor, α, measured in dB m^{-1}. Final calculation of transmission loss, TL, is effected by applying the formula

$$TL = k \log r + \alpha r$$

where k is chosen as 20 for free-field, 10 for cylindrical or 15 for "practical" spreading. For purposes of computer calculation, with f in kHz, the following formulae apply to the calculation of α, to an accuracy acceptable for sonar calculations

$$\alpha = \alpha_1 + \alpha_2 + \alpha_3$$

$\alpha_1 = af^2$ \qquad (Freshwater attenuation, [2.2])

$\alpha_2 = bf_0 (1+(f_0/f)^2)^{-1}$ \qquad (MgSO$_4$ relaxation, [2.3])

$\alpha_3 = cf_1(1+(f_1/f)^2)^{-1}$ \qquad (Boric acid relaxation, [2.4])

where, below, S is salinity (‰) and T is temperature (° C). Also

$a = 2.1 \times 10^{-10} (T - 38)^2 + 1.3 \times 10^{-7}$

$b = 2S \times 10^{-5}$

$f_0 = 50(T+1)$

$c = 1.2 \times 10^{-4}$

$f_1 = 10^{(T-4)/100}$

At 35‰ and 14° C check values are $a = 2.5 \times 10^{-7}$, $b = 7.0 \times 10^{-4}$, $f_0 = 750$ kHz and $f_1 = 1.26$ kHz.

Planktonic marine life, suspended material and entrained gas bubbles can all lead to "anomalously" higher attenuation than might be predicted by the equations discussed above. These various entities contribute to attenuation, if present in sufficiently great densities, by a combination of scattering and resonance effects. The latter phenomenon is most likely to be evident in bubbly water, or where plankton which may contain or which may respire gas bubbles is encountered. All such effects are difficult to quantify. In general if field experiments suggest a level of attenuation greater than can be ascribed by invoking the standard transmission loss equations, experimental error should be the first assumption, not anomalous attenuation. An excellent review of the subject of attenuation of sound in sea-water is provided in reference [2.4].

Attenuation in marine sediments [2.5] is another area of potential importance in some sonar situations. The attenuation of sound in marine sediments has been found to vary approximately exponentially with both distance and frequency. It is specified by an attenuation coefficient, α_s, measured in dB m^{-1} which is, itself, a function of frequency given by

$$\alpha_s = \beta f^\gamma$$

Experimental measurements indicate that γ is close to unity for marine sediments. The total attenuation, in decibels, for sound which has travelled a distance r within a single sediment layer and which has a transmission frequency f, is thus

$$\alpha_s r = \beta f r$$

The attenuation coefficient α_s is, itself, a function of **bottom porosity**, n, which is defined as

$$n = V_w/V$$

where V_w is the volume of water in a sample of sediment and V is the volume of the sediment.

Alternatively, **moisture content**, m, may be quoted and is taken to be the weight of water in a sample, divided by the weight of solids, so that

$$n = (\rho_s/\rho_w)(m/(m + 1))$$

where ρ_s is the sediment density and ρ_w is the density of water.

Table 2.1
Marine sediment properties

	n	ρ_s (kg m^{-3})	m	c (m s^{-1})	σ (MRayl)	β
Coarse sand	0.4	2000	0.25	1800	3.6	0.3-0.6
Fine sand	0.5	1900	0.35	1700	3.2	0.4-0.7
Silt	0.6	1800	0.50	1600	2.9	0.1-0.5
Clay	0.8	1400	1.33	1600	2.2	0.05-0.2

Marine sediments may become heavily saturated with hydrocarbon gases, as a result of discharges from deeper gas and petroleum reservoirs, or as a result of biological decomposition. Saturation with gas will cause the sediment to become highly reflective to sound waves. Normal attenuation levels will then be far exceeded.

2.5 Target Strength

Target strength, TS, is the decibel measure of intensity returned from a target, referred to 1 m standard range from the notional "acoustic centre" of the target, relative to the incident intensity. Suppose for example, that an active sonar is used to gauge the target strength of a midwater target, such as a "trials" submarine, located vertically below the source vessel, figure 2.2. Imagine the depth of the submarine to be 120 m, the sonar projector electrical power input to be 500 W and the echo return intensity to be 143 dB re 1 µPa. The calculation proceeds as follows. The source level is evaluated, as we have seen in section 2.2, as

$$SL = 10\log P + 167 = 194 \text{ dB re 1 µPa}$$

To calculate transmission loss, we assume free-field conditions in deep water and hence spherical spreading, so that

$$TL = 20\log r = 20\log 120 = 42 \text{ dB}$$

The intensity arriving at the submarine is

$$SL - TL = 194 - 42 = 152 \text{ dB re 1 µPa}$$

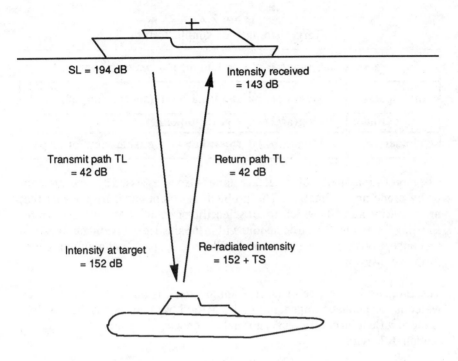

Figure 2.2 An example of the use of the sonar equations in assessing target strength

We presume the submarine to have a target strength TS. The signal strength at one metre notional range is then 152 + TS. We insert the return transmission loss, to find the intensity at the surface vessel. This we have measured at 143 dB re 1 µPa. Simplifying and solving for target strength, we find that

$$TS = 33 \text{ dB re 1 } \mu Pa$$

The example quoted above gives an indication of the way in which target strength may be measured. For military targets, such as submarines, mines and torpedoes, such measurement is common practice. Optical assessment of target highlights used to be carried out by photographing a model of a particular type of submarine, painted matt black. Only particularly good reflective planes "stood out" against a matt black background when the model was illuminated with bright light. A more modern approach utilises images of the submarine created by computer graphics techniques, to achieve the same end. A theoretical assessment of target strength can be made for some of the simpler solid forms, some of the commoner of which are listed in table 2.2.

Table 2.2
Target strength of simple forms

a = radius; range = r; L = length; k = wavenumber		
Solid sphere	$10 \log(a^2/4)$	r>>a; ka >> 1 (sphere "large")
Long cylinder	$10 \log(ar/2)$	e.g. oil-pipeline
Cylinder	$10 \log(aL^2/2)$	r>>a; ka>>1 e.g. side view of torpedo

The target strength of biological targets can also be assessed by measurement or by crude approximation. The problem is complicated by the fact that many marine animals which are of scientific or commercial interest exhibit shoaling behaviour. Social behaviour within the shoal may invalidate statistical preconceptions regarding member distribution in space and may thus cause further confusion.

The acoustic impedance of marine animal flesh is about 1 MRayl. Since water has an acoustic impedance of 1.5 MRayl, we may assess the absolute value of reflection coefficient at normal incidence (using our earlier formula, section 1.7, with $\theta_1 = 0$) as

$$|R| = (\sigma_2 - \sigma_1) / (\sigma_2 + \sigma_1) = 0.2$$

The attenuation of sound in animal tissue is high, by comparison with attenuation in water, being of the order of decibels **per centimetre**, rather than per metre. We see that marine organisms are thus only moderate reflectors of sound.

Some planktonic organisms, and all teleost fish, have one characteristic which can lead to unusually high echo returns, however. This feature is the presence of a gas "bubble" within the animal (the "swim-bladder" in the case of fish). Gas bubbles, unlike rigid reflecting spheres, exhibit high target strength at frequencies corresponding to their mechanical "resonance" frequency. Because shoaling organisms do not return a "coherent" reflection (as might the sea-bed, were it suitably smooth) the shoal acoustic characteristics are better described in terms of scattering rather than reflection coefficients.

2.6 Reflection Intensity Loss Coefficient

When sound reflects in a **specular** manner (that is, mirror-like, without scattering) from a smooth surface, the intensity reflection coefficient may be defined as

$$\mu_r = 10\log(I_3/I_1)$$

where I_3 and I_1 are the intensities following and preceding reflection, as

$$I_3 = \overline{p_3^2(t)}/\sigma_1 \qquad I_1 = \overline{p_1^2(t)}/\sigma_1$$

in our previous figure 1.5. We have that
So that the intensity loss coefficient in
decibels is
$$\mu_r = 10\log(\overline{p_3^2}/\overline{p_1^2}) \equiv 20\log(|R_{12}|)$$

where R_{12} is the pressure reflection coefficient defined in section 1.7.

2.7 Sea-floor Loss

Those parts of the sea-floor which are sedimentary deposits are often capable of being regarded as a sensibly plane, smooth, reflector of sound. Acoustic impedance is then dominant in determining acoustic properties. Combining the above results with those presented in section 1.9 we establish the **Rayleigh** bottom-loss reflection model for reflection at the interface between media themselves, presumed lossless.

Of course, marine sediments must be, to some extent, lossy because the passage of pressure waves through them causes particle motion and, through friction, the dissipation of acoustic energy. To take account of the fact that sediments are lossy, we have the **NUC** (Naval Underwater Center) model [2.6], which is empirical and specified in terms of bottom porosity, $0 < n < 1$, which gives bottom-loss directly in decibels

$$-(17.5n-1.025)f^{1/3}[\tanh((6.55 - 0.0724\theta_1)n)^{(1.5/n)} + (0.08-0.296n)(1 - 0.0117\theta_1)^2]$$

Sample curves are shown in figure 2.3. Here **f** (in bold) is the transmission frequency in kHz, θ_1 is the angle of incidence measured in degrees, relative to the normal, as before, and the bottom porosity is as defined in section 2.4 above.

A useful and adequately accurate approximation for the tanh function in the NUC model is

$$\tanh\xi = \xi(1.09 - \xi(0.395 - 0.0472\xi)) \text{ for } \xi < 3$$
$$= 0.995 \text{ for } \xi > 3$$

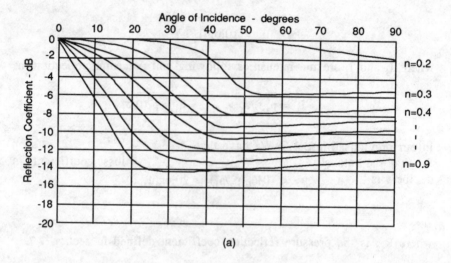

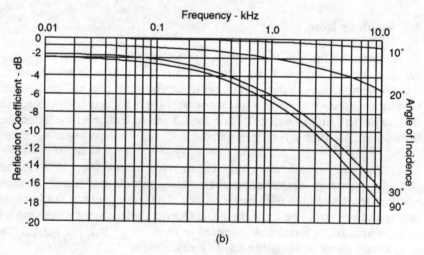

Figure 2.3 The NUC bottom loss model: (a) versus angle of incidence and bottom porosity at a transmission frequency of 1 kHz and (b) versus frequency and angle of incidence at a bottom porosity of 0.5

2.8 Sea-surface Loss

The sea-surface acts both as a reflector and as a scatterer. When calm, reflection with no scattering occurs. Then, as we have already seen in section 1.8, because the impedance of air is vanishingly small, by comparison with that of water, $\sigma_2 = 0$ so that $A = 0$ and $R_{12} = -1$. Also $T_{12} = 0$ for all angles of incidence. Total internal reflection therefore always occurs.

The Sonar Equations

When the sea-surface is rough, however, scattering as well as specular reflection takes place and any of several empirical models may be used to predict reflectivity. The criterion governing the action of the sea-surface as a reflector is thus its surface roughness, which is determined by the **Rayleigh parameter**

$$Q = k\sigma\cos\theta$$

where k is the sound wave-number ($k = \omega/c$), $\boldsymbol{\sigma}$ (written in bold to distinguish it from specific acoustic impedance) is the root-mean-square crest-to-trough wave-height (the standard deviation of the wave-height distribution function) and θ the angle of incidence, measured to the normal to the sea-surface. The surface pressure reflection coefficient is given as $R = -\exp(-Q)$ and this result is considered to apply when $Q \ll 1$. If $Q \gg 1$, then the surface acts primarily as a scatterer, dissipating the incident radiation in a non-coherent manner.

The following equation inter-relates wave height σ and windspeed, w, for a fully arisen sea

$$\sigma = 0.005w^{2.5}$$

The **Beckmann–Spizzichino** surface reflection loss model [2.7] permits calculation of the reduction of acoustic intensity following reflection, internally within the ocean, from the sea-surface. The surface reflection coefficient is a function of angle of incidence to the horizontal, θ (herein measured in degrees), windspeed, w (knots) and frequency, f (kHz).

The geometry of the problem is defined as follows. Sound of acoustic intensity I_1 is incident upon the sea surface from below, at an angle θ to the normal to the surface. That sound is in part specularly reflected, with intensity I_2, and in part scattered and absorbed. The surface loss coefficient is defined, in decibels, to be the logarithmic ratio

$$\mu_r = 10\log_{10}(I_2/I_1)$$

A convenient algorithm for calculating the Beckmann–Spizzichino loss coefficient in decibels is presented here as

$$\mu_r = 10\log_{10}((1 + (f/f_1)^2)/(1 + (f/f_2)^2)) - (1 + (90 - w)/60)((90 - \theta)/30)^2$$

where $f_1 = \sqrt{10f_2}$ and $f_2 = 378w^{-2}$. "Spot check" parameter and loss values are $w = 20$, $\theta = 60$, $f = 7$, loss = -11.518.

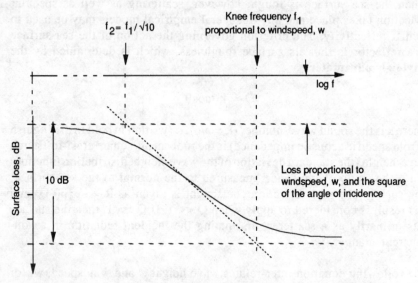

Figure 2.4 The Beckmann–Spizzichino loss coefficient

Presenting the loss algorithm in this form has the advantage that the interaction of the various parameters governing loss may be easily visualised. On log-frequency, log-attentuation (decibel) scales, figure 2.4, the "flattened Z" of the loss curve is fixed in shape for all conditions of parameters. It will slide leftwards with increasing windspeed, indicating increasing attentuation at yet lower frequencies as the sea-surface becomes more agitated. Under this same condition, the entire curve will also shift downwards, reflecting a gross overall increase in attenuation. Finally, the curve will exhibit a gross downwards shift for increasing angle of incidence, again as might be expected from a heuristic argument.

2.9 Noise

We shall consider the extensive subject of noise in the sea more fully in chapter 6. For the purpose of illustrating the necessary manipulations, in the context of the sonar equations we note two principal classifications of acoustic noise which may corrupt a received signal: **ambient noise,** generated by a variety of natural and man-made mechanisms, and **self-noise,** produced by hull friction, the wake and bow-wave, and machinery on board a vessel carrying a listening sonar. For the moment we concentrate upon the more general principles of noise analysis and commence by observing that noise must be specified as a spectral density.

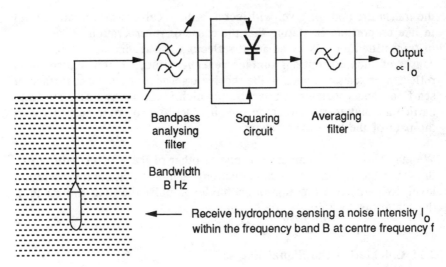

Figure 2.5 Analysis equipment for measurement of noise spectral density

In order to measure a noise spectrum, we utilise some form of analyser such as is shown in figure 2.5. The tunable measuring filter will exhibit a bandwidth of B Hz at a frequency, f, and the entire system will be so calibrated as to yield a measure of received intensity within this bandwidth, again in decibels relative to our reference unit. To plot the spectrum, we plot dB re 1 μPa per Hz of measurement filter bandwidth. That is, if the analyser output at frequency f is I_o dB re 1 μPa, then we plot $N(f) = I_o - 10 \log B$ dB re 1 μPa per Hz. To utilise such a spectral plot, we need to know the useful bandwidth – W, say – of our sonar, as well as its transmission frequency. The calculation may then be reversed to yield a corrupting noise level

$$NL = N(f) + 10 \log W$$

2.10 Reverberation

As with noise, we cover this topic more fully in chapter 6. We note however that reverberation, like noise, is a corrupting influence which may mask sonar returns. We note also that, unlike noise, reverberation is caused by the operation of the sonar, being the result of the reflection of transmitted signals back towards the receiver by adjacent scatterers or surfaces. Whereas increasing signal power will assist in improving signal detectability against noise, this will not be the case with reverberation. In this case, increasing the transmitted power level will increase the received reverberation level in like proportion. In stating that the cause of reverberation is the return of transmitted signals by adjacent reflectors, we recognise two principal causes of such corruption: **surface reverberation,** wherein the sound is reflected or scattered back to the receive transducer by the sea-surface or sea-floor, and **volume reverberation,** which is caused by reflection from particulate matter, marine organisms, bubbles and so on, in suspension in the path of the transmitted beam.

We may anticipate dominance of one or other of these two conditions in the vast majority of operational situations, and note that the reverberation level, RL, will be a function of source level, pulse duration and range to the scattering location.

2.11 Calculating the Signal Excess

The ultimate objective in applying the sonar equations is to determine the level of detectability of received signals. Many sonars are **monostatic**: the transmit and receive transducers are at essentially the same location. **Bistatic** sonars would utilise transmit and receive transducers at widely differing locations. For echo-ranging purposes such systems are uncommon, although the bistatic geometry is inherent in communication, navigation and positioning systems. For a monostatic, echo-ranging system, we should seek to determine a signal excess as

$$SL + 2DI - 2TL + TS - NL$$

If the sonar is reverberation-limited, rather than noise-limited, replace NL by RL. As with the distinction between the two possible causes of reverberation, it is usually the case that *either* noise *or* reverberation will present the dominating corrupting influence.

For a bistatic sonar, such as an acoustic communication link, we need only consider the single-way transmission loss, and target strength is, of course, irrelevant. We evaluate the signal excess (essentially the signal to noise ratio) as

$$SL + DI_t + DI_r - TL - NL$$

where DI_t and DI_r are the transmitter and receiver transducer (or array) directivity indexes, respectively. Most often, for reasonably efficient communication, the signal excess would need to be of the order of 10 dB. For echo-ranging, and some positioning systems, a much worse signal excess can be tolerated, either because signal repetition can be used to integrate the signal out of the noise or because the display format itself may offer substantial visual integration, as with the XYt plots common in sub-bottom profiling and sidescan sonar work.

References

[2.1] L. Hall, The Origin of Ultrasonic Absorption in Water, *Phys. Rev.*, Vol. 73, 1948, p. 775

[2.2] L.N. Liebermann, Origin of Sound Absorption in Water and in Sea Water, *J. Acoust. Soc. Am.*, Vol. 20, 1946, p. 868

[2.3] E. Yeager et al., Origin of the Low Frequency Sound Absorption in Sea Water, *J. Acoust. Soc. Am.*, Vol. 53, 1973, p. 1705

[2.4] R. Urick, *Sound Propagation in the Sea,* Peninsula Publishing, Los Altos, Calif., 1982, pp.5.1-5.17

[2.5] E.L. Hamilton, The Elastic Properties of Marine Sediments, *J. Geophys. Res.*, Vol. 76, 1971, pp. 579-604

[2.6] H.R. Hall and W.H. Watson, An Empirical Bottom Reflection Expression for Use in Sonar Range Prediction, *NUC Technical Note 10*, July 1967

[2.7] R. Coates, An Empirical Formula for Computing the Beckmann–Spizzichino Surface Reflection Loss Coefficient, *IEEE Trans. Ferroelectrics Frequency Control*, Vol. 35, No. 4, July 1988, pp.522-3

3 Characteristics and Analysis of Sonar Waveforms

3.1 Introduction

Sonar emissions and the noises which corrupt them are pressure waves travelling in a four-dimensional space–time continuum. In the past, it has often been adequate to restrict consideration of such processes only to the time-domain signals emerging from the outputs of each of the hydrophones used for signal detection. Increasing computing power, together with a rapid evolution in appreciation of the mathematics of signal analysis and its application to particular physical problems must now lead the underwater acoustician towards a more all-embracing comprehension of the spatio-temporal nature of the processes he is called upon to handle.

In this chapter, it is not intended to cover each possible method of characterisation or analysis in extreme detail. Copious references are provided to allow the reader to delve deeper into any particular method which may commend itself as being particularly appropriate for a given task. Rather, the objective is to provide a chart of a complex and often confusing territory which will, hopefully, provide some insights into method and applicability for the wide range of interesting and powerful analysis techniques and equipments which have become readily available to the marine acoustician.

Although popularly described as "the silent world", the sea is, as we might by now suspect, an extremely noisy environment. Sound propagates well, at sonic frequencies, and derives from sources which are many and varied. Thus far, our characterisation of the sound has been in terms of source level alone. That is, we have been interested only in average power output. Common sense would suggest that we should seek a deeper insight into the temporal, spectral, spatial and directional characteristics of the man-made signals received by sonar equipments and also of the marine sounds which may corrupt or mask them.

Temporal behaviour has to do with the way in which source level varies, as a function of time, in the short term. We might expect that some types of sound source, for example noise caused by distant storms, would maintain some constancy of power level over quite long periods (longer at least than our inspection period). On the other hand, "vocalisation" by marine animals would exhibit relatively rapid changes in power level. Sound sources of the first kind would be referred to as stationary, because their average power level is constant on a time-scale which is relatively long by comparison with the "period" of their fastest amplitude fluctuations. Those in the second category are **non-stationary** and would be expected to exhibit relatively fast fluctuations of short-term averaged power.

Both sorts of waveform, as exemplified above, would be further classified as **finite power processes**. This allows us to distinguish them from burst-like emanations, such as explosive detonations which, although admittedly non-stationary, differ from the continuous, if fluctuating, sounds produced by marine animals in that they are of limited duration. They are **finite energy processes**. It should be stressed that the boundaries between many of these classifications are blurred when an attempt is made to use them to characterise perceived marine sounds. For example, the crackle produced by myriad snapping shrimps is a sound source produced by marine animals, and is almost always best described as a stationary, finite power process.

Spectral behaviour is concerned with the distribution in frequency of the power or energy content of a waveform. For stationary, finite power processes, we examine a long-term averaged **power spectral density**, which plots power per Hertz of measuring system bandwidth versus frequency. For non-stationary, finite energy processes, we attempt to establish an **energy spectral density**. Finally, for non-stationary, finite power processes, we have recourse to an **energy time–frequency plot**. Clearly these various forms of representation require further explanation.

We shall also consider the **temporal correlation** attributes of signals. That is, any two signals may share some similarity of waveform as, for example, when a signal is subject to transport delay (a common enough occurrence with sonar waveforms). The source and received signals may well not be exactly the same because, given sufficient transport delay, noise will have been added, and attenuation and other corrupting influences will have been at work. The temporal correlation function, the **cross-correlation**, will indicate both the extent of the delay and the degree of comparability between the two waveforms. Indeed, the method also works with a single waveform, to unravel, for example, contained echoes. The correlation function is then referred to as an **auto-correlation**.

The estimation of a temporal correlation function may be described as a transformation from the time-domain into the **delay-domain**. It is not the only such transformation. As we shall see, evaluation of the correlation function of a signal can be effected by (amongst other techniques) Fourier Transformation of the power or energy spectral density (as appropriate) for that signal. If, instead, we Fourier Transform the **logarithm** of the power spectrum, we obtain a new delay domain characteristic, referred to as the **cepstrum**. The term cepstrum is simply an anagram of the word "spectrum". The word has, of itself, no deeper significance than that.

The reader will probably be aware that, in evaluating a power spectral density or an energy spectral density, phase information that would be present in, for example, the Fourier Transform (the voltage spectral density) will be lost. However if, in evaluating the logarithm of the power spectrum, phase information from the calculation of the voltage spectral density is retained, and used to establish a complex log power spectrum, then transformation into the delay domain yields a complex cepstrum, which may be processed to eliminate or suppress chosen bands of delay wherein, for example, reverberant information might lie. By reversing the procedure and transforming back to the time domain it is possible to establish signal waveforms free from contamination by echoes. This process is referred to as **homomorphic deconvolution.**

In establishing a temporal correlation function we compare two waveforms subject to a relative time displacement between them. Alternatively, we may assume both processes to be recorded with, as it were, zero time slippage, but with a variable spatial displacement. This allows us to compute a **spatial correlation function**. The spatial correlation function is related by a transform identity, to the **angular distribution of acoustic intensity** of the sound field.

3.2 Swept Frequency (Heterodyne) Spectrum Analysers

The most basic of the commonly available spectrum analysers operates by shifting the baseband spectrum of the waveform being analysed to a higher frequency, by mixing with a local oscillator, the output of which is swept in frequency. The baseband spectrum, thus translated, is slid across a narrow-band bandpass analysing filter. The filter is followed by a precision envelope detector to yield an oscilloscope display Y signal. The X signal is provided by the ramp waveform inducing the frequency sweep. The analysis procedure, illustrated in figure 3.1, is analogous to the superheterodyne principle used in radio receivers, to obtain high selectivity following radio-frequency tuning which is a coarse selection process with poor rejection of out-of-band noise and interference. Superheterodyne detection involves using a multiple stage, intermediate frequency amplifier, with each stage a double-tuned, critically-coupled radio-frequency transformer. Attempting to sweep the centre-frequency tuning on such an amplifier would be quite difficult. Consequently, for practical swept-frequency analysers, sweeping the centre-frequency of the analysing filter would only be contemplated in relatively low-frequency (10 kHz or less) specialised applications. Much useful advice on the practical design of swept spectrum analysers, including selection of filter characteristics, sweep rate and averaging times for both deterministic and random signals is to be found in reference [3.1].

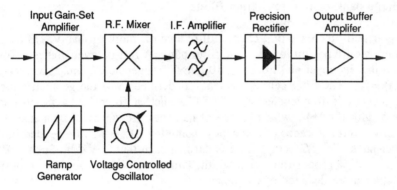

Figure 3.1 *The heterodyned frequency sweeping spectrum analyser*

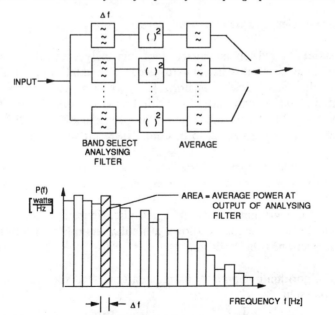

Figure 3.2 *The filter-bank analyser and the interpretation of analyser output power spectrum*

3.3 Filter-bank Spectrum Analysers

A different approach to spectrum analyser design is to be found in the filter-bank analyser, which is illustrated in figure 3.2. Here, the need to sweep is eliminated. A virtually instantaneous appraisal of the entirety of the spectrum is obtained and the possibility exists for effecting a display which will adequately mirror spectral change when examining a quasi-stationary process. However, it is arguable that resolution will be potentially inferior to that of the swept analyser, simply because of the cost of implementing a large bank of fine resolution filters.

It is usually the case that either full-octave or, more probably, 1/3 octave filters would be employed in the filter bank. A full-octave filter is defined such that its upper and lower −3 dB cutoff frequencies, f_u and f_l, are related as $f_u = 2f_l$. The filter centre frequency is defined to be the geometric mean of the cutoff frequencies: $f_c = (f_u f_l)^{1/2}$. Such a filter has a proportional bandwidth $(f_u - f_l)/f_c$ which is 70% of its centre frequency. This means that, to cover three decades of frequency, some ten filters will be needed in the filter-bank. The 1/3 octave filter is defined such that $f_u = 2^{1/3} f_l$. Such a filter offers a 23% proportional bandwidth. Thirty such filters would be required to span three decades of frequency.

3.4 Fast Fourier Transform Analysers

The most flexible and sophisticated of spectrum analysers currently available, the "FFT" analysers, make use of hardware implementations of an algorithm known as the Fast Fourier Transform [3.2] which is, itself, a particularly efficient method of computing a result known as the Discrete Fourier Transform (DFT). An introduction to the properties of the DFT and the nature and use of the FFT is to be found in reference [3.3].

The DFT calculates Fourier Series components from a sample sequence segmented from the process to be analysed. The segment is presumed to represent one period of a hypothetically periodic waveform. Because the input waveform will not in general be periodic, figure 3.3(a), averaging on a segment by segment basis allows a stable, smoothed spectrum of an aperiodic fuction to be built up.

The DFT algorithm computes N harmonically related spectral coefficients $X(n)$ as the sum

$$X(n) = \frac{1}{N} \sum_{k=0}^{N-1} x(k) \exp(-2\pi jkn/N) \qquad n = 0, 1 \ldots N-1$$

Characteristics and Analysis of Sonar Waveforms

where – as figure 3.3(b) illustrates – x(k) represents the sampled, periodic input vector derived from a bandlimited time function v(t) containing a highest frequency f_o, by sampling at equal time intervals $T = 1/f_o$, so that

$$x(k) = v(kT); \quad k = 0, 1 \ldots N-1$$

An inversion transformation also exists and can also be computed using the FFT algorithm. This is the Inverse Discrete Fourier Transform (IDFT). The IDFT has the form

$$x(k) = \sum_{n=0}^{N-1} X(n) \exp(2\pi jkn/N) \quad k = 0, 1 \ldots N-1$$

The inter-relation between the (line) spectrum and the input data sequence is shown in figure 3.3(a) and (b).

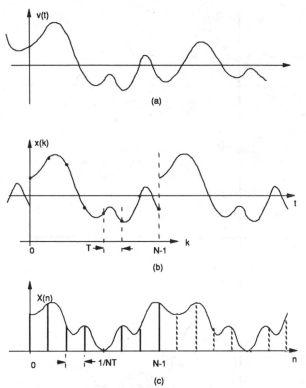

Figure 3.3 *Illustrating the waveform sampling and spectral periodicity inherent in the Discrete Fourier Transform, applied to a segmented aperiodic, bandlimited waveform sampled in accordance with the Sampling Theorem*

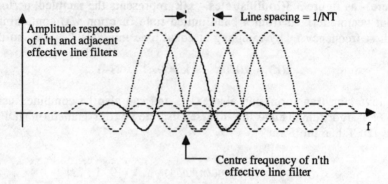

Figure 3.4 The representation of the DFT as being equivalent to a "sincx" transfer-function parallel filter-bank

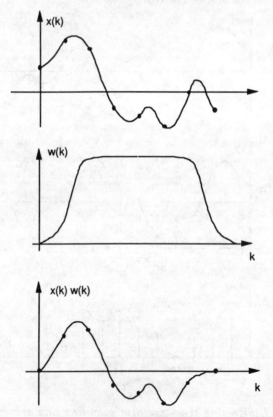

Figure 3.5 Windowing to reduce terminal discontinuities and reduce leakage between spectral lines by modification of the "sinx/x" filter transfer function induced by rectangular windowing

A time-series such as defines each DFT spectral coefficient can be taken to represent, also, a Finite Impulse Response (FIR) digital filter. The FFT analyser is thus, in this respect, directly analogous to the filter-bank analyser, except that its operation is numerical, rather than analog-electronic. By employing the methods of Laplace or z-transform calculus, it is possible to elucidate the effective transfer function of each "filter" in the DFT "filter-bank". The filters, if the DFT is used in its most primitive formulation (that is, literally as stated above) are poorly selective by comparison with well-designed analog octave or 1/3 octave filters such as would be used in an analog filter-bank analyser.

Each DFT "line filter" has a "sin x/x" response against frequency, figure 3.4. This means that, if a large spectral line lies partway between adjacent spectral line locations, leakage into the adjacent lines (and, indeed, other close-to lines) will occur. This problem can be ameliorated by impressing a windowing function [3.4] upon the abstracted segment as figure 3.5 shows.

Because the windowing process "loses" some information, the segmentation process will probably now be engineered to include some degree of overlap, thus potentially reducing data throughput and hence curtailing effective analyser input bandwidth.

Because the FFT analyser relies on a hardware implementation which is firmware-controlled and software-evolved, great flexibility in processing options can be made available to the user. In particular and quite unlike the swept-filter and filter-bank analysers, the FFT analysers can compute (subject to sampling limitations, of course) literal Fourier Transformations of input waveforms. This means that they may display not only power or energy spectra, but the amplitude and phase functions of a suitably synchronised input waveform. Yet more sophisticated processing is possible. The analyser, having computed a power spectrum, may transform again, to display a correlation function. The majority of FFT analysers will allow input of two waveforms, simultaneously, making possible the calculation of cross-spectra and cross-correlation functions.

Finally, the ability to synchronise FFT analysers, in "grabbing" a waveform segment, means that the analyser may be triggered by, for example, an internal combustion engine firing cycle. This makes possible most sophisticated displays of phenomena which are only short-term stationary, such as would derive from the detonations which take place within the engine.

3.5 Prony Analysis [3.5]

The DFT approach to spectral analysis remains the workhorse of waveform analysis techniques. However, it is possible to call upon other procedures which may, under some circumstances have advantages. The DFT equation, in one sense, presumes a structure to the waveform being analysed. For example, suppose we were to employ the technique in analysing a sound from a musical instrument which was rich in harmonics. Suppose also that we could synchronise the sampling period to that of the fundamental of the musical instrument. Then the DFT "filter bins" would co-locate with the harmonics of the sound, and great precision in identification of amplitude and phasing of the harmonics would result.

The generation and need for identification of tonal components in a waveform is by no means unusual in underwater acoustics. For example, the self-noise of a ship will contain readily identifiable line spectral components. They will derive from the whines, whistles and hums associated with reciprocating and rotating machinery. There is no prior reason to assume that such tonal components will exhibit harmonic inter-relationships although such a feature may from time to time be present. For example, blade-rate radiation from a three-bladed propellor will occur at the third overtone of shaft vibrations. By contrast, where there is gearing between an engine and a shaft, the gearing ratio will not, in all probability, be a simple integer relationship and tonal components initially deriving from a single source will not exhibit harmonic relationship.

Furthermore, in using the DFT, it is common to employ a large number (typically 512 or 1024) of spectral samples. There may be far fewer tonal components of profound interest in, say, a ship's self-noise spectrum. An alternative analysis option due, with much subsequent modification, to Prony and thus bearing his name involves modelling the signal (or perhaps more accurately the signal generating mechanism) as a suite of anharmonically related sinusoidal generators, the amplitudes, phases and frequencies of which are to be adjusted by a least squares procedure to minimise an error vector between the model time series, $x_m(k)$ and the signal time series $x(k)$. Thus we might write

$$x_m(k) = \sum_{n=0}^{N-1} A(n) \cos(\omega(n) t + \phi(n)) \qquad k = 0, 1 \ldots K-1$$

and attempt to minimise $\{x_m(k) - x(k)\}^2$. The result of the minimisation is an array of values of the vector $[A, \omega, \phi]$ for the N values of n. Drawing the spectrum then means simply drawing in zero-width lines which purport to locate, and specify amplitude and phase of, the anharmonic components of $x(k)$.

Of course, the method will work only poorly if x(k) is rich in atonal spectral components; that is, if the spectrum is dominated by noise-like sounds. This could be the case (at least in part) when ship sounds are being analysed, because the bow-wave and propellor cavitation give rise to such processes. Clearly, the choice, use and interpretation of these various analysis techniques require care and experience. Finally, it should be noted that the technique described above is predominantly employed in an off-line, software-dominated context although, with the ready availability of highly portable computing machinery of great power and flexibility, no major difficulty would be thought to attend a hardware, real-time implementation.

3.6 Further Model-building Techniques for Spectral Estimation [3.6, 3.7]

The Prony method is a subset of a more general class of model-building methods of analysis which have assumed great popularity during the past decade. This wider class of techniques is based upon the assumption that many sound-generating mechanisms can be modelled as, effectively, noise-stimulated digital filters. The general form of a digital filter, specified in the frequency domain, is as a z-operator [3.8] transfer function

$$H(z) = B(z)/A(z)$$

where $z = \exp(-j\omega T)$ and T is the sampling interval. A(z) and B(z) are polynomials capable of representation in the factored form $A(z) = (z + a_1)(z + a_2) \ldots (z + a_m)$; $B(z) = (z + b_1)(z + b_2) \ldots (z + b_n)$. This amounts to supposing that a sound generating mechanism may be at least identified by a z-plane pole-zero filter model and that even its very physical structure may yield to such an interpretation. In some ways, such a line of thought should not, perhaps, be surprising. Many physical structures capable of vibrating are representable as interacting collections of lumped energy storage and dissipative elements. Thus the practical application of the theory of differential equations and the use of transform methods in their solution should be recalled as a commonplace, in the investigation of such phenomena.

Whether the sampled-data or difference equation form is truly likely to be innately representative of underlying physical structure is, perhaps, to be doubted. Its utility in the context of digital computation is clearly evident although it should be remembered that, with the current rapid increase in computing power, the rather simple integration model implied by the z-transform operator may yet be superseded by digital processing which more closely parallels the end-effect of the analytical process of integration and thus permits a return to a more direct attack upon modelling the actual physical structure of the generating mechanism.

Use of the model presented above in a practical spectral analysis context can take on any of several different forms. If the A(z) = 1, then the model is essentially that of a moving-average (MA) filter and the analysis is referred to in that way. If the B(z) = 1, the model is an all-pole digital filter and is referred to as being auto-regressive (AR). If both A(z) and B(z) are used to model the process, then it is referred to as an auto-regressive, moving-average (ARMA) analysis.

The objective of the analysis procedure is the moving about of the pole/zero locations on the z-plane so as to minimise the mean-square error between some appropriate attribute of the filter output and some chosen equivalent attribute of the waveform being matched. This could involve, paralleling in a sense the Prony method described in the previous section, the minimisation of the error between the filter output vector and a vector containing the sampled signal being analysed. In fact, the analysis takes as its input a (short) Blackmann–Tukey [3.9] time-domain derived **correlation function** and uses the ARMA, or MA or AR technique, to establish a matching correlation function in the filter output. Since the correlation function is a statistical attribute of the digital filter, rather than a deterministic output, a white noise excitation of the filter input (numerically) would be entirely satisfactory.

It may seem strange at first sight, particularly to those who are familiar with the speed tradeoff in computing correlation functions, using an FFT algorithm by first calculating a power spectrum, then re-transforming. However, it is appropriate to recall several features of the application of ARMA-type modelling. First the model size or order will be modest by comparison with the size of an FFT. Then the method may yield attractive improvements in the quality of a spectral estimate, by comparison with the FFT approach, particularly if record lengths *are* short.

Finally, we should ask how does the method produce a spectral estimate, if all that has happened is the jiggling of pole/zero positions, to match up target waveform and model output waveform *correlation* functions? Notice that, once the jiggling process has been completed to within some required level of error magnitude, we are left with a transfer function H(z) defined in terms of pole and zero (or pole, or zero) locations. By re-writing H(z) with z replaced by exp(–jωT), we obtain a function in ω, which may then be readily reduced to the form of a power spectrum, since

$$P(\omega) = |H(\exp(-j\omega T))|^2$$

3.7 Four-dimensional Space–Time Waveform Analysis

Imagine, as figure 3.6 suggests, a model of the ocean, wherein exists a spatially distributed pressure field which is the result of the combined effect of a multiplicity of sound sources. In this illustration one might be tempted to suppose that a single source existed somewhere down in the bottom rear left-hand corner. This could be the case, but such a visualisation actually simplifies the reality, where wavefronts from many directions would interweave to create a locally fluctuating pressure at a hydrophone location. What possible observables and statistical measures might interest us?

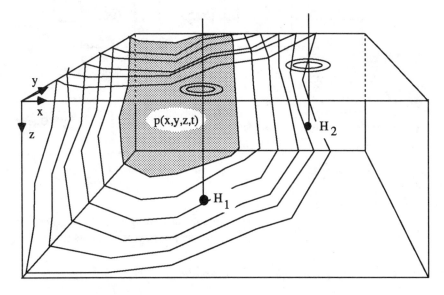

Figure 3.6 A randomly fluctuating sound-field with two sensing hydrophones H_1 and H_2

At hydrophone H_1 at co-ordinates (x_1,y_1,z_1) we might observe pressure $p_1(t)$ = $p(x_1,y_1,z_1,t)$. In fact, the hydrophone output would be a voltage $v_1(t)$ = $k_h p_1(t)$, where k_h is the hydrophone constant, measured in volts per μPa. Since it is probably easier for most readers to think in terms of the hydrophone output voltage and its various attributes, when manipulating units, than the actual pressure field itself, we shall continue by utilising this form of the sound field variable. We can, of course, think immediately in terms of a Fourier Transform of $v_1(t)$: $V_1(f) \Leftrightarrow v_1(t)$. However, it is probably better to skip this stage of thinking and to anticipate other frequency-domain, time-frequency and delay-domain attributes of the process.

We thus observe $v_1(t)$, the hydrophone output voltage which is proportional to the time-domain instantaneous pressure fluctuation and which is measured in volts [V]. We may also seek to determine, by calculation or direct measurement, one of the following spectral quantities, because each allows us to form an appreciation of the distribution of power or energy (as is deemed appropriate) within the waveform as a function of frequency, or of both time and frequency. An electronic mechanism for achieving these objectives is shown in figure 3.7.

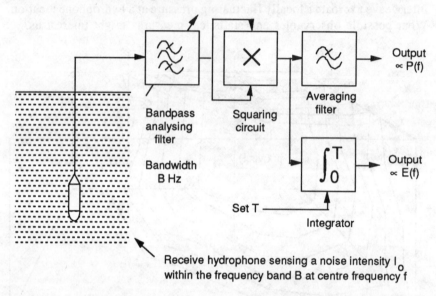

Figure 3.7 *The measurement of power and of energy density estimates by means of a narrow-band, band-pass filter*

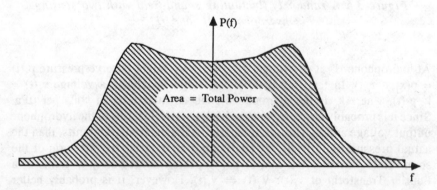

Figure 3.8 *Integration of power spectral density to yield total power*

$P_1(f)$: the **power spectral density**, measured in [$V^2 Hz^{-1}$] or, preferably, [$V^2 s$], assuming $p_1(t)$ to have been a stationary, finite power process, or

$E_1(f)$: the **energy spectral density**, measured in [$V^2 s\, Hz^{-1}$] or, preferably, [$V^2 s^2$], assuming $p_1(t)$ to have been a non-stationary, finite energy process, or

$e_1(t,f)$: the **short-term energy spectral density**, measured in [$V^2 s^2$], assuming $p_1(t)$ to have been a quasi-stationary, finite energy or finite power process.

Notice, figure 3.8, that the integral of $P_1(f)$ with respect to frequency yields power – the average power in a continuously transmitted broadband signal, perhaps. The integral of $E_1(f)$ with respect to frequency yields energy – the total energy in an explosive detonation, for example. The integral of $e_1(t,f)$ versus time and frequency (the volume beneath the $e_1(t,f)$ surface over the t,f plane) also yields total energy, the function itself depicting, figure 3.9, energy observable within a short epoch Δt by an analysing filter of width Δf.

Let us pursue the quest for observables deriving from $v_1(t)$ further. We may attempt to observe a correlation function $R_1(\tau) = \int v_1(t) v_1(t + \tau)\, dt$ which seeks to measure the internal temporal similarity within $v_1(t)$. That is, if we skip over a time interval τ, do we observe some measure of similarly between the delayed version of v_1, namely $v_1(t + \tau)$ and $v_1(t)$ itself? For example, if v_1 were a periodic function, of period T, then we would expect to encounter strong correlation at delay intervals $\tau = nT, n = 0, 1, 2 \ldots$. Notice that correlation may be observed in both finite power and finite energy signals. It is not uncommon, for example, to observe strong correlation in the (finite energy) waveform following an explosive detonation because, particularly in shallow water, a high level of reverberation will be present, and delayed and attenuated replicas of the explosive signature will be contained within the hydrophone output waveform. We note that the correlation function, as expressed above, actually pertains only to finite energy processes. For finite power processes, the integral would necessarily become infinite, in the limit. We should write, more correctly

$$R_{11}(\tau) = \lim_{T \to 0} \frac{1}{T} \int_{-T/2}^{+T/2} v_1(t) v_1(t+\tau)\, dt$$

for finite power processes and

$$R_{11}(\tau) = \int_{-T/2}^{+T/2} v_1(t) v_1(t+\tau)\, dt$$

for finite energy processes. The "11" subscript denotes **auto-correlation** estimation and is introduced here in anticipation of the need to define a delay-domain "cross-correlation", as between the outputs of hydrophones H_1 and H_2. Note that the forms of correlation coefficient for the finite power and finite energy processes are intrinsically different, dimensionally. The reader will probably be familiar with the Wiener-Kinchine relationships

$$R_{11}(\tau) \Leftrightarrow E_{11}(f)$$

for finite energy processes and

$$R_{11}(\tau) \Leftrightarrow P_{11}(f)$$

for finite power processes, with the appropriate correlation integral being used to define the relationship between R_{11} and v_1 in each case. We further note that it is often the case that a normalised and dimensionless correlation function will be employed, by computing a quantity $R1_1(\tau)/R_{11}(0)$. This quantity has extreme values of ± 1.

Thus far we have concentrated on observables associated with a single point of measurement (x_1, y_1, z_1). At two different points (x_1, y_1, z_1) and (x_2, y_2, z_2) we might expect to encounter some usefully interpretable element of delay-domain correlation as, for example, when the second point of observation lies further back along a line drawn from an acoustic source through the first point of observation. This is just the same as using two hydrophones

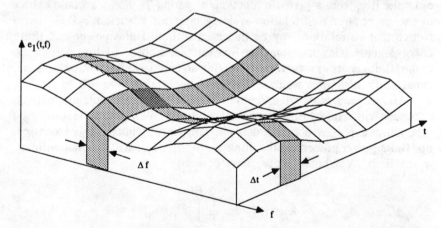

Figure 3.9 The e(t,f) spectrum, showing that the volume beneath the surface yields a measure of the total energy in the signal within an epoch Δt and over a frequency band of width Δf

to form a crude array, which might be pointed at a source to determine azimuth and bearing angle. The important quantity needed in interpreting the hydrophone outputs is the baseline distance between them which, together with sound-speed, allows the differential delay to be determined. This delay will be greatest when the two hydrophones are co-linear with the source. We thus seek to measure or calculate a cross-correlation $R_{12}(\tau)$ where

$$R_{12}(\tau) = \lim_{T \to 0} \frac{1}{T} \int_{-T/2}^{+T/2} v_1(t) v_2(t + \tau) dt$$

for finite power processes and

$$R_{12}(\tau) = \int_{-T/2}^{+T/2} v_1(t) v_2(t + \tau) dt$$

for finite energy processes. These quantities are related to cross-power, $P_{12}(f)$ and cross-energy $E_{12}(f)$ spectral densities, respectively. They describe a similarity of power or energy spectral components between the two locations as, for example, when directional hydrophones observe similar attributes of a signal, at different spatial locations, when looking at a source, but may observe different attributes of a directionally variable sound or reverberation field corrupting the source signal. The relationships (again chosen to be appropriate for the type of signal being investigated) are $R_{12}(\tau) \Leftrightarrow P_{12}(f)$ and $R_{12}(\tau) \Leftrightarrow E_{12}(f)$.

It was mentioned in section 3.1 that the cepstrum [3.10] provides another delay-domain method of examining a waveform. In order to appreciate better the computational steps involved in cepstral processing we shall concentrate upon finite power functions so that, given a hydrophone output $v(t)$ we may hope to obtain a power spectrum $P(f)$, from which we establish the power cepstrum as $c(\tau) \Leftrightarrow \log P(f)$. Why should this apparently trivial processing step be of any value in waveform analysis? The reason is perhaps most easily understood and, from our standpoint as underwater acousticians, is best exemplified by considering the problem of multipath propagation (or equivalently, reverberation caused by echoes from the sea-surface and sea-floor). For this purpose, we imagine that the cepstrum is to be used in a diagnostic sense, to tell us something about the reverberation delay: the difference in arrival time between a main-path signal and one or more echoes.

If we represent the main path signal, as registered at the hydrophone output, as $v(t)$ then a multipath signal may be represented as $k(v(t + \tau_o))$, where k is a loss-factor caused by the increased path length of a reflected signal as well as loss on reflection, and τ_o represents the differential delay between the two paths, which we wish to be able to estimate. The hydrophone output is then

$$v_o(t) = v(t) + k(v(t + \tau_o)) \Leftrightarrow V(f) + k(V(f)\exp(2\pi jf\tau_o))$$
$$= V(f)(1 + k\exp(2pjf\tau_o))$$

Manipulating the right-hand side of the transformation, we note that the spectrum at the hydrophone output is that of the transmitted signal, but with an impressed, sinusoidal modulation so that

$$|V_o(f)|^2 = |V(f)|^2 \{(1 + k\cos(2\pi f\tau_o))^2 + k^2\sin^2(2\pi f\tau_o)\}$$
$$\propto |V(f)|^2 \{1 + k\cos(2\pi f\tau_o)\}$$

Now, frequently, the source signal $V(f)$ will be of "low-pass" form, exhibiting decreasing spectral components with increasing frequency. This means that the power spectrum of $v_o(t)$ will have the form shown in figure 3.10(a). Transformation into the delay-domain, as a correlation function, allows the sinusoidal ripple to transform as a more or less well identified spike located at delay magnitude τ_o, as figure 3.10(b) shows. Of course, if the delay magnitude is relatively small, the spectral ripple period will be large and

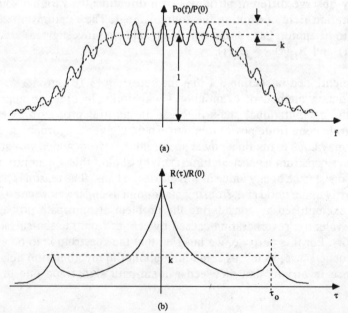

Figure 3.10 The power spectrum contaminated with a ripple modulation in frequency caused by multipath, and the corresponding correlation function, showing the delay identifying spikes

very few cycles of ripple will be available for the IDFT analysis to work on. Identification and use of the delay-spike will then become difficult. If, however, the logarithm of the power spectrum is calculated, then the ripple amplitude remains large over the entire frequency range, even at low delay magnitudes. Consequently, the cepstrum may be thought to provide a preferable mechanism for delay-attribute identification and measurement than, say, a correlation function.

In fact, because the non-linear operation implied by the logarithmic function adds no further information to the signal being processed, it is without doubt the case that further clever processing would cause a correlation function to yield as good information concerning delay. The cepstrum should, then, really reside as a convenient, rather than a unique weapon in our armoury in analysing signals which have passed through a reverberant channel. Indeed, because the logarithmic operation on the spectrum distorts a sinusoidal ripple-shape into a cusp ripple-shape, transformation produces not a single delay-domain spike, but a periodic multiplicity of such spikes. With more than one delay path present, this can make the cepstrum at least as difficult to interpret as an auto-correlation function.

As it happens, an ingenious variant of the cepstrum, known as the meta-cepstrum [3.11], may get around this difficulty. Instead of applying a logarithmic function generation, imagine the numerical equivalent of a variable-gain amplifier being passed across the spectrum, almost to establish spectral whitening by automatic gain control in the frequency domain. Then – considerably more closely, at least – the ripple will retain its sinusoidal form as well as maintaining constant amplitude across the full span of the spectrum.

It was also mentioned in section 3.1 that a delay-domain suppression of echoes within $v(t)$ may be possible. To do this we form a complex cepstrum. At this point we must make special note of a small point of philosophy, concerning the use of the cepstrum method. We would normally imagine that sampled data computations, to evaluate such attributes as spectra and correlation functions, are but convenient or, indeed, in the sphere of digital computation, necessary approximations to otherwise impractical "analog-world" transformations which would be presumed to act upon continuous signals.

In the case of the cepstrum – and particularly in the case of the complex cepstrum – this is not really true. The complex cepstrum is essentially an attribute of the sampled signal, rather than of the signal itself. This is because, in order to form the complex cepstrum, we have no alternative but to segment our time-series representing $v(t) \Leftrightarrow v(k); k = 0, 1 \ldots N\text{-}1$

and, using a DFT, transform to compute an unsmoothed spectrum estimate $V(n)$; $n = 0, 1 \ldots N-1$. We note that, as calculated by the DFT, the transform $V(n)$ will consist of a real and imaginary part: $V(n) = X(n) + jY(n)$ from which amplitude is calculable as $A(n) = (X^2(n) + Y^2(n))^{1/2}$ and phase is calculable as $\phi(n) = \tan^{-1}(Y(n)/X(n))$. We then form the complex quantity $\ln(V(n)) = \ln(A(n)\exp(j\phi(n))) = \ln(A(n)) + j\phi(n)$ and apply the IDFT to retransform this quantity into the delay-domain.

Thus far, we have investigated the temporal, spectral and delay-domain properties of the sound field. Although, in calculating cross-correlation functions, we happen to have investigated a spatially variable geometry, we have not explicitly sought a statistical measure of similarity which was actually a function of, let us suggest, a displacement vector

$$\mathbf{d}_{12} = (x_1, y_1, z_1), (x_2, y_2, z_2)$$

That is, a correlation function which would seek to establish a statistical average such as

$$E\{p(\mathbf{d}_{01})\, p^*(\mathbf{d}_{01} + \mathbf{d}_{12})\}$$

By way of example, let us restrict the problem to one wherein we seek to determine the **spatial correlation** resulting from measuring the cross-power spectral density

$$C(\zeta) = E\{V_1(\omega)V_2^*(\omega)\}$$

where V_1 and V_2 are Fourier Transforms of the hydrophone outputs. That is, for a single separation, ζ, we move the vertical pair to depth z and measure the hydrophone cross-power spectral density. We move to a new depth and repeat, averaging in order to assemble by this means the function C for fixed separation. By repeating again for different separations, we build up a picture of the spatial function itself. The spatial cross-power spectral density may be Fourier Transformed to yield a new function, referred to as a **wavenumber spectrum,** $Q(p)$ [3.12]. We write

$$C(\zeta) = C^*(-\zeta)$$

and calculate

$$Q(p) = \frac{1}{2\pi} \int_{-\infty}^{+\infty} C(\zeta) \exp(-ip\zeta)\, d\zeta$$

In this case, because the vertical hydrophone pair cannot respond to variability in the horizontal plane, we must presume that (as for example in an open ocean situation) no innate horizontal directivity is to be encountered. The reader interested in pursuing further the question of angular directivity and spatial correlation is directed towards reference [3.13]

References

[3.1] R.B. Randall, *Frequency Analysis*, Bruel & Kjaer, Naerum, DK-2850 Denmark, 1987 [ISBN 87 87355 07 8]

[3.2] J.W. Cooley and J.W. Tukey, An Algorithm for the Machine Computation of Complex Fourier Series, *Math. Comp.*, Vol. 19, 1965, pp. 297-301

[3.3] R. Coates, Fourier Transform Methods, in *An Introduction to Digital Filtering* (Bogner, R.E. and A.G. Constantinides, ed.), Wiley, 1975

[3.4] F.J. Harris, On the Use of Windows for Harmonic Analysis with the Discrete Fourier Transfor", *Proc. IEEE*, Vol. 66, Jan. 1978, pp. 51-83

[3.5] S.L. Marple, Spectral Line Analysis by Pisarenko and Prony Methods, *IEEE Conf. Acoustics, Speech and Signal Processing*, 1979, pp. 159-161

[3.6] S.M. Kay and S.L. Marple, Spectrum Analysis – A Modern Perspective, *Proc. IEEE*, Vol. 69, No. 11, 1981, pp.1380-1419

[3.7] O.L. Frost, Power Spectrum Estimation, in *Aspects of Signal Processing* (G. Tacconi, ed.), Reidel Publishing, Dordrecht, The Netherlands, 1977, pp. 125-162

[3.8] A.V. Oppenheim and R.W. Schafer, *Digital Signal Processing*, Prentice-Hall, New Jersey, 1975

[3.9] R.B. Blackmann and J.W. Tukey, *The Measurement of Power Spectra from the Point of View of Communications Engineering*, Dover, New York, 1959

[3.10] D.G. Childers, D.P. Skinner and R.C. Kemrait, The Cepstrum: A Guide to Processing, *Proc. IEEE*, Vol. 65, No. 10, 1977, pp. 1428-1443

[3.11] P. Hirsch, The Metacepstrum, *J. Acoust. Soc. Am.*, Vol. 69, No. 3, 1981, pp. 863-865

[3.12] H. Cox, Spatial Correlation in Arbitrary Noise Fields With Application to Ambient Sea Noise, *J. Acoust. Soc. Am.*, Vol. 54, 1973, pp. 1289-1301

[3.13] S.M. Morfey and C.L. Baxter, *Angular Distribution Analysis in Acoustics*, Springer-Verlag, Berlin, 1986 [Series: Lecture Notes in Engineering, No. 17]

4 Ray Trace Modelling of Sonar Propagation

4.1 Introduction

Sonar modelling has to do with predicting sound intensity at some point in the sea remote from a source. It provides a more detailed way of predicting performance than do the sonar equations, which would usually be used as a "first cut" and preferably "worst-case" approach to system design. Sonar modelling is of great importance in deducing the path traversed by sound as, for example, in seismics – where it is required to determine the thickness and acoustic characteristics of sea-bed sediment layers – or in military applications where range and bearing to an underwater sound source, such as an enemy submarine, must be found. A comprehensive review of modelling software currently in use is to be found in reference [4.1].

Most frequently, the modelling problem is defined by assuming a known vertical sound velocity profile as characterising a channel of fixed depth. Horizontal variability of sound velocity is often discounted, on the grounds that it is variable only over distances substantially greater than typical ocean sonar ranges. This assumption should be regarded as dubious in shelf-sea, or estuarine conditions or when operating at extremely low frequencies. The structure of, and nomenclature associated with the world seas and oceans is shown in figure 4.1. Table 4.1 provides an indication of the dimensions associated with the various submarine features illustrated in this diagram.

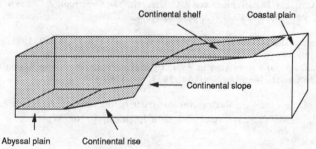

Figure 4.1 The structure and nomenclature of the ocean

Table 4.1
Typical dimensions of various ocean features

	Av. depth (m)	Max. depth (m)	Area x10^6 (m^2)	Sound speed characteristics
Atlantic Ocean	3900	9100	82	Ocean: Fig. 4.10
Pacific Ocean	4300	11000	165	Ocean: Fig. 4.10
Arctic Ocean	1200	4600	14	Isothermal, 0°C
North Sea	94	700	0.6	Isothermal, 2–10°C (seasonal)
Mediterranean Sea	1400	5100	3	Ocean: Fig. 4.10; S:38 ppt (av.)

Topography of continental margins

	Width (km)	Gradient
Shelf	75 (av.)	0.0002 (av.)
Slope	20–100	0.005 (av.)
Rise	0–600	0.0001–0.001

Modelling takes as its most fundamental basis, some attempt to solve the wave equations, approximately and typically numerically. Two principal methods exist, with several variants which we shall not discuss. The first method parallels the processes of geometric optics, assumes a horizontally stratified medium and is referred to as **Ray Tracing.** The second method, known as the **Mode Theory** approach, develops particular solutions to the wave equations, analytically, which describe the ability of the channel to enter preferred "resonant states". Numerical methods are then used to compute pressure at a specified depth, and range as a function of time and frequency. The use of Mode Theory thus avoids the necessity, incumbent upon the Ray Tracing method, of establishing the entire ensemble of rays at all points between source and receiver. Ray Tracing is suitable for applications where sound wavelength is small by comparison with range and water depth. Mode Theory is often considered complementary to Ray Tracing, being suitable for application to shallow water channels. It is a subject that we shall examine in greater depth in Chapter 5.

4.2 Ray Tracing Sonar Models [4.2–4.5]

Ray tracing involves the application of Snell's law to a horizontally stratified medium. **The most important concept in ray tracing is that of the ray**

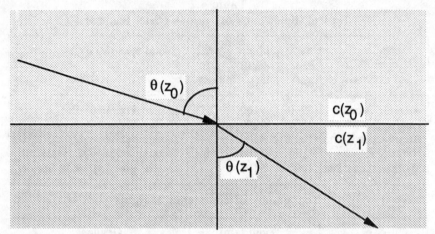

Figure 4.2 Refraction through water layers of differing sound speed

coefficient "a". If a ray is launched at angle $\theta(z)$ from a source at depth z, figure 4.2, then on intersecting a boundary between media of differing acoustic impedance

$$\sin(\theta(z_o))/\sin(\theta(z_1)) = c(z_o)/c(z_1)$$

If many layers are present, figure 4.3, then this result may be generalised in the following way:

$$\sin(\theta(z_o))/c(z_o) = \sin(\theta(z_1))/c(z_1) = \sin(\theta(z_n))/c(z_n) = \mathbf{a}$$

and the parameter **a** is seen to depend on the initial launch angle and the initial sound speed, **and to remain a constant for that ray thereafter**. By applying this principle of constancy of ray coefficient along any given ray, computer programs may be written to plot the courses of typical rays launched from a source. The programs may also compute distance travelled along a ray, and may be used to infer intensity at points remote from the source.

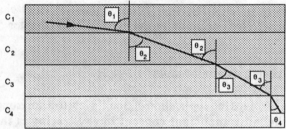

Figure 4.3 Refraction through many layers: constancy of the ray parameter a

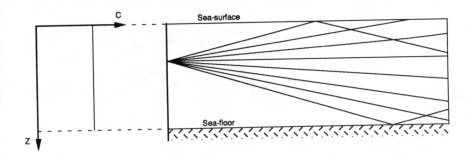

Figure 4.4 Propagation in isovelocity (constant velocity with depth) conditions

In order to be able to understand how ray tracing software evolves, we look next at the way in which a ray will pass through a layer of constant sound speed. That is, **isovelocity** conditions pertain, with $c(z) = 1500$ m s^{-1} for all depths. Then propagation takes place in straight line paths, as figure 4.4 illustrates. Suppose, then, that we return to our stratified model, and imagine that sound speed is constant within each layer. Then, as figure 4.5 shows, we imagine the sound speed profile with depth to be a histogram-like approximation to what, in reality, would be a continuous and smoothly changing sound speed profile. Now, if a ray is launched at depth z_0 in the first layer, at an angle θ_0 to the horizontal, one may compute first the distance to the second layer, and thus obtain end-point co-ordinates for plotting the first part of the ray, and then proceed on downwards through the remaining layers to compute the entire ray trajectory. Clearly, the finer the gradation of the stratification, the smoother will become the computed ray trajectory. Equally clearly, so also will computation time increase.

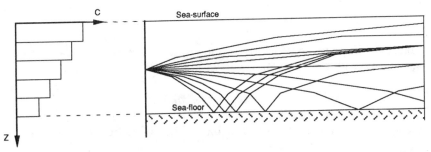

Figure 4.5 Stepwise, stratified approximation to a sound speed profile decreasing linearly with increasing depth

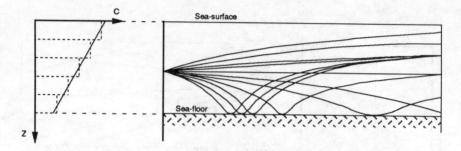

Figure 4.6 Using linear interpolation of sound speed within the stratified layers

An obvious step in improving matters might involve considering an alternative approximation to the sound speed versus depth profile. Consider, for example, a situation in which we assume sound speed to vary linearly, across a layer after the manner shown in figure 4.6. It can then be shown that sound rays travel in circular paths of specifiable radius and centre co-ordinates. Now the computations become more complex, but the curves so generated are smoother and fewer layer approximations become necessary. Whilst one might contemplate going to even higher orders of approximation – making for example, parabolic fits to the sound speed profile – this has not been considered worthwhile in the past.

4.3 Ray Trace Calculations

For the linear sound-speed fit within a layer, the important results governing ray path calculations are summarised here. It should be stressed that much greater detail can be built into a ray-trace program, which can become an entity of quite considerable complexity. The following equations will allow the interested student to begin a process of development of modelling routines suitable for use on any of a wide range of fast, graphics-oriented modern microcomputers.

Figure 4.7 depicts the geometry of the problem. For a ray, of ray constant **a**, moving within a layer of water with a sound speed profile

$$c(z) = z_0 + bz$$

the sound will travel in a circular locus [4.6] of radius, r, given by

$$r = (ab)^{-1}$$

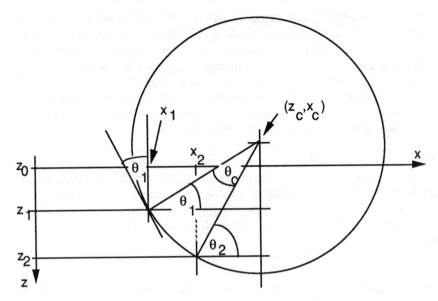

Figure 4.7 Defining the geometry of circular ray propagation within the depth layer z_1 to z_2

If the starting co-ordinates for the ray are (z_1, r_1) then the centre of the upward-curving circular locus will have co-ordinates

$$(z_c, x_c) = ((z_1 - r \sin \theta_1), (x_1 + r \cos \theta_1))$$

Notice that the centre of the downward curving locus, encountered when the sound speed gradient in a stratum is negative rather than positive, will have co-ordinates

$$(z_c, x_c) = ((z_1 + r \sin \theta_1), (x_1 + r \cos \theta_1))$$

In either event, the equation of the circle so defined is

$$(z - z_c)^2 + (x - x_c)^2 = r^2$$

so that, if the approached bound of the stratified layer in which the ray is moving is at depth z_2, then the terminating co-ordinates for the arc will be (z_2, x_2) where x_2 will be given by

$$x_2 = x_c + 0.5 \{2x_c^2 - (z_2 - z_c)^2 + r^2\}^{1/2}$$

Since many modern microcomputers will access graphics software (or even, in some instances, hardware) capable of effecting circular arc plots directly and with great speed, the equations given above, plus some control logic, afford the basis for a simple ray tracing program. Care is needed to trap the ray turning situation and to cope with the isovelocity condition b=0, which corresponds to straight line propagation and by implication an infinite radius for the circular locus. Logic is also needed to determine the intercept with the sea-floor and initiate the reflected ray.

Further equations which allow computation of distance and approximate travel time along a ray segment are given below. t_1 is taken to be the time at the point of entry of the ray to the layer, at location (z_1,x_1). t_2 is the time of arrival at location (z_2,x_2). The length along the arc is determined by evaluating first the angle to the vertical, θ_2 at (z_2,x_2).

$$\theta_2 = \sin^{-1}(a\ c(z_2))$$

The included angle at the centre of the circle is then $\theta_c = \theta_2 - \theta_1$, and the length of the arc is simply $\delta s = r\theta_c$. The average sound speed within the stratum is given by

$$\bar{c} = c_0 + b(z_1 + z_2)/2$$

Consequently the approximate travel time is given as

$$\delta t = \delta s / \bar{c}$$

More general expressions are to be found in the literature [4.2] for travel time along a ray. However, since starting data for sound speed profiles, for example, is often but poorly known, some considerable insight into sound propagation may yet be acquired with models based even upon these simple rules.

4.4 Some Examples of Ray Modelling

In this section, we shall assume a simplified statement of sound speed as a function of depth, temperature and salinity which recognises approximate **coefficients of sound speed** for these variables of 0.0016 m s^{-1} per metre increase in depth, 4.6 m s^{-1} per Centigrade degree increase in temperature and 1.3 m s^{-1} per part per thousand increase in salinity, giving

$$c = 1450 + 4.6T + 0.0016z + 1.3(S - 35)$$

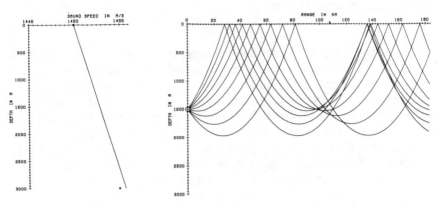

Figure 4.8 Sound rays propagating in upward circular arcs in isothermal Arctic water

Isothermal water, such as might be found in the Arctic Ocean or in any deep ocean below the main thermocline, leads to a sound speed profile which rises approximately linearly with depth, because it is the increase of sound speed with increasing pressure that becomes the dominant effect. Thus in Arctic water with T = 0 °C and S taken as 35 ppt:

$$c = 1450 + 0.0016z$$

The effect of projecting rays from a source is shown in figure 4.8. The upward circular ray paths are readily identified. It will also be noticed that some rays cannot reach the ocean-floor. The turning depth for a ray launched at depth z_1 and at a downward angle $\theta(z_1)$ to the vertical is readily calculated. The ray constant for such a ray is

$$a = \sin \theta(z_1)/c(z_1) = \sin \theta(z_t)/c(z_t)$$

where z_t is the turning depth. But at the turning depth $\sin \theta(z_t)$ must be unity since $\theta(z_t)$ will be 90°. Therefore we find that

$$c(z_t) = c(z_1)/\sin \theta(z_1)$$

and since $c(z_t)$ is expressible, for isothermal water, as an equation of the form

$$c_0 + bz_t$$

where c_0 is the surface sound speed and b the pressure coefficient with depth, it follows immediately that

$$z_t = b^{-1}(c(z_1)/\sin \theta(z_1) - c_0)$$

Water in the **main thermocline** decreases in temperature, linearly with depth. In the deep ocean, the thermocline might extend downwards to a depth of, perhaps 500 m exhibiting a temperature fall of 10°C through this interval. The temperature coefficient with depth is thus –0.02°C per metre increase in depth. In the thermocline, with S again taken as 35 ppt, the temperature-induced fall in sound speed with increase in depth is (about) –0.02 x 4.6 = 0.092 ≈ 0.1 m s^{-1} per metre increase in depth, which is nearly two orders of magnitude larger than the pressure coefficient of 0.0016 m s^{-1} per metre. The sound speed equation thus becomes

$$c = 1450 - 0.1z; \quad 0 < z < 500$$

This sound speed profile gives rise to downward circular ray paths within the thermocline, as figure 4.9 shows. Below the thermocline will exist isothermal deep-ocean water at, perhaps, 4° C. Consequently, from the base of the thermocline at 500 m depth, to the ocean-floor at 2500 m, the sound speed profile is (as with our first example) dominated by the pressure

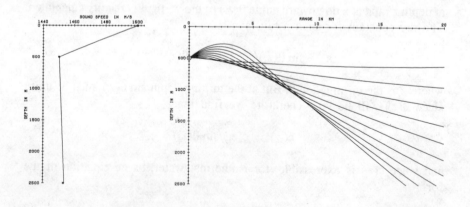

Figure 4.9 Sound rays propagating in downward circular arcs within the main thermocline

Ray Trace Modelling of Sonar Propagation

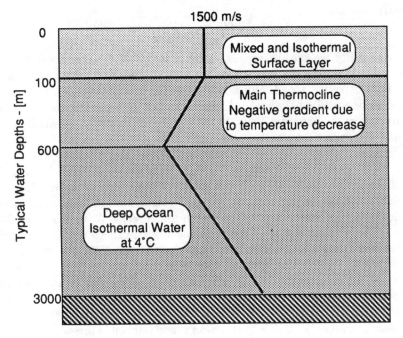

Figure 4.10 Idealised deep ocean sound speed profile

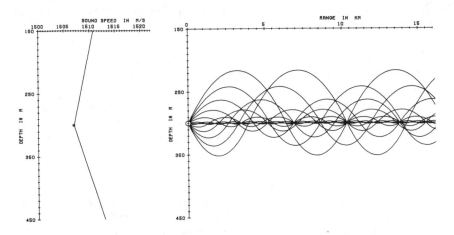

Figure 4.11 Sound trapped in an acoustic waveguide: projector at lower boundary of main thermocline

coefficient and
$$c = 1400 + 0.0016z; \quad 500 < z < 2500$$

Consequently we see the rays, as they pass out of the thermocline enter an upward-curving circular locus, but now (because of the much smaller gradient of sound-speed with depth) with far less pronounced curvature. Also to be noted on figure 4.9 is the formation of a **caustic** on the lower boundary of the pattern of rays. In the vicinity of a caustic, unusually high sound intensities may be encountered.

A typical ocean sound speed profile, such as is shown in figure 4.10, adds to these various possibilities a surface layer some tens of metres deep, in which wave action induces mixing and establishes roughly isovelocity conditions, for which $z = 0$, $S = 35$ ppt and $T = 14°$ C, so that $c = 1514$ m s^{-1}. The peculiar nature of the deep ocean sound speed profile results in some interesting transmission phenomena. As the ray trace diagram shown in figure 4.11 illustrates, a sound source on the lower surface of the main thermocline radiates rays which are trapped in its vicinity by the increase in sound speed which occurs with increasing displacement from it. Such sound channels are believed to be used by whale pods in establishing trans-oceanic communication. This is because sound trapped in this way is subject only to cylindrical spreading loss and, at sonic frequencies, suffers a relatively small attenuation loss. It is interesting to note, also, that the Blue Whale has been discovered to emit sounds at a source level equivalent to that of a modern warship. This makes it not only the largest but also the loudest animal species.

The ray diagram shown in figure 4.12 also illustrates the formation of the sound channel, when the source is located at the base of the thermocline, where the sound speed gradient changes sign. All operating and propagation conditions remain as in figure 4.11, except that the angular spread of rays from the source has been increased significantly. The plotted area is also increased, to show reflections from the sea-surface. The strong channelling along the base of the thermocline is clearly in evidence.

Finally, some operating and propagation conditions may introduce divergences such that **shadow zones** of unusually low insonification are formed, figure 4.13. It is not true that no sound at all exists within shadow zones. Diffraction effects will still produce weak insonification. Also, regions of high sound intensity, called **focii**, may be observed. Notice that the only difference between the operating and propagation conditions between this figure and figure 4.12 is that, in the former, the sound source was located at the base of the thermocline at a depth of 500 m. Here it is only 50 m below the surface. The effect on the ray-plot of this change in depth is, however, profound.

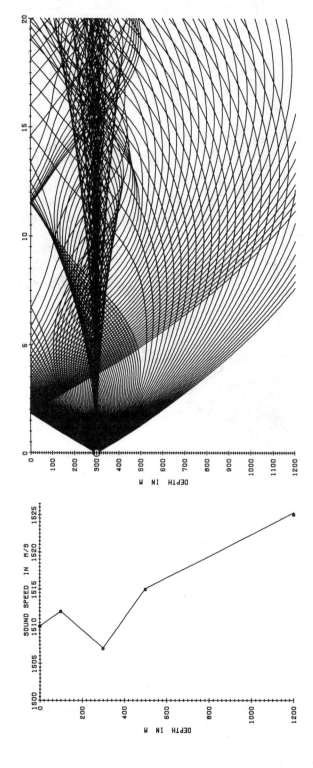

Figure 4.12. Showing the formation of the sound channel when the source is located at the base of the thermocline

64 Underwater Acoustic Systems

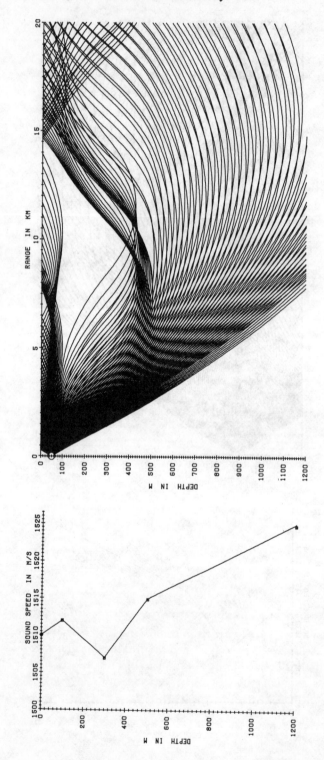

Figure 4.13. Showing the formation of focii and shadow zones

4.5 Modelling Transmission in the Shelf-seas

The shelf-seas are those shallow seas which surround the continents but which form the continental shelf. Typically, they will be only some tens of metres deep. The North Sea, for example, is a shallow shelf sea with an average depth of 94 m but which over large parts of its area is less than half that depth. During the major part of the year, weather conditions and tidal currents establish good mixing and lead to isovelocity conditions and an absence of a marked thermocline. Water temperature, however, will vary markedly both over the region and throughout the year.

To place these observations in context, a sequence of North Sea velocity profiles was analysed to establish an average variation of sound speed, which indicated a slight increase with depth, caused primarily by increase in temperature, and at the level of 0.05 m s^{-1} per metre. This figure, although small, still significantly exceeds the pressure coefficient of sound speed, as we have assumed it above, of 0.0016 m s^{-1} per metre. Its significance may be further appreciated if we note that the radius of curvature of sound rays will be calculated as at least 30,000 m, this figure applying when a ray is launched horizontally, thereby producing greatest curvature. It is clear that, over kilometre ranges, propagation will be essentially in straight lines.

Propagation at ranges which considerably exceed water depth places some significant computational demands upon the modelling procedure. In essence, the sound channel then consists of an isovelocity medium separated by plane, parallel reflecting surfaces. The reflection coefficient of these surfaces will, of course, depend upon environmental conditions, frequency and angle of incidence. For the moment, however, let us assume perfect reflection and proceed by considering an optical analogy.

Imagine, figure 4.14(a), that a object, O, and a point of inspection, P, are defined, within the region separating plane, parallel mirror surfaces. There will be a direct path between the object and point of inspection. There will also, figure 4.14(b), be a pair of reflections, providing first images I_{11} and I_{12} in mirrors 1 and 2 respectively. The notation I_{m1} and I_{m2} is used to signify the m'th multiple reflection in these mirrors.

Outside of mirrors 1 and 2 there will exist multiply reflected images of these mirrors and figure 4.14(c) illustrates the first such pair of images of mirrors 1 and 2. As seen from point P, there will now appear image I_{21}, the result of apparent reflection in the first image of mirror 2. Similarly, there will also appear an image I_{22}, because of reflection of image I_{12} in the first image of mirror 1.

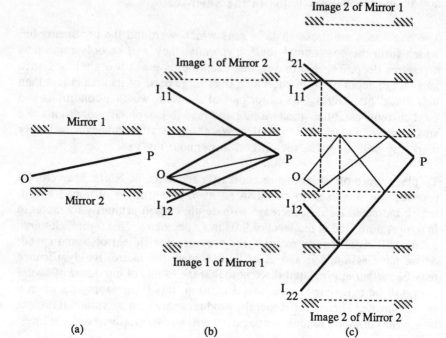

Figure 4.14 *Illustrating the formation of multiple images of a source by two plane parallel reflecting surfaces*

Clearly, this pattern will repeat, ad infinitum, with I_{21} reflecting in the second image of mirror 1, I_{22} reflecting in the second image of mirror 2 and so on. In acoustical terms, the result of this process of sustained multiple reflections will be from the observer's viewpoint at P, an effective array of periodically placed sources. In the acoustic case, mirror 1 will be the sea-surface, offering phase inversion on reflection. Mirror 2 will be the sea-floor, which we shall presume to be a hard reflecting boundary offering no phase inversion. We can build the phase functions into general pressure reflection coefficients, the choice of which is largely at the discretion of the investigator.

For example, we might under some circumstances choose to employ the sea-bed reflection models giving a bottom reflection coefficient $R_b(\theta) = R_{12}$, with R_{12} as specified in sections 1.9 to 1.11. R_{12} is, of course, a function of incident angle, θ, as well as the sea-floor physical properties. Similarly a simple model for the sea-surface reflection coefficient might just be $R_s(\theta) = -1$; all θ. Alternatively, we might introduce more sophisticated models, developed from, perhaps, the sea-floor and sea-surface intensity reflection models presented in sections 2.7 and 2.8 remembering, of course, that such models do not inherently build in phase information and present intensity rather than pressure coefficients which are directly calculated in decibels rather than as a numerical ratio. Our procedure will follow the geometry developed in figure 4.15.

In this figure we see, on the left-hand side, the first few reflection-pairs in the sea-surface, the sea-floor and the multiple reflections of both. The source is placed high in the water, at depth z_1, so that its first reflection in the sea-surface is relatively close to it, and its first reflection in the sea-floor is more remote. The pairing of reflections is then quite marked. If the reader examines the paths taken by rays passing from the image sources to the receiver location, each will be seen to pass through a certain number of surface and bottom reflection layers. Passage through a surface reflection layer (each is marked "S" on the diagram) will correspond to multiplication of the acoustic pressure at the layer by R_s, the surface reflection coefficient. Similarly, passage through a bottom reflection layer (marked "B") will cause the pressure to be multiplied by the bottom reflection coefficient R_b.

The reader should also note that reflection image pairs are identified by number. That is, we identify pairs m = 1, 2, 3 ... above and below the true water layer. The "m = 1" pair "above" actually includes the source itself and is thus partially within the true water layer, of course.

Notice also, that we denote the ray-length, in a given ray pair, for that ray which is closer to the horizontal axis of the true water layer, as r_{m1} and the ray-length corresponding to that which is further away, as r_{m2}. Finally, we prime and double-prime the r's to denote ray-pairs which start, respectively, "above" or "below" the true water channel. There are thus, for any one value of m, four ray-lengths to consider in estimating spreading loss and transport delay between the image sources and the receiver location: r'_{m1} and r'_{m2} as well as r''_{m1} and r''_{m2}.

If we inspect the right-hand diagram, which shows the m'th ray pair below the true water channel, we observe that

$$r''_{m1} = (((2mh - z_2) - z_1)^2 + R^2)^{1/2}$$

$$r''_{m2} = (((2mh - z_2) + z_1)^2 + R^2)^{1/2}$$

If we examine the m'th ray pair above the true water channel, we find that

$$r'_{m1} = (((2mh + z_2) - z_1)^2 + R^2)^{1/2}$$

$$r'_{m2} = (((2mh + z_2) + z_1)^2 + R^2)^{1/2}$$

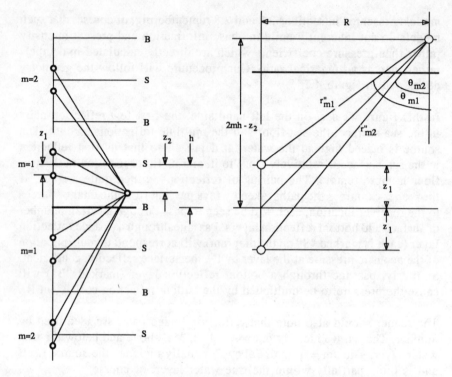

Figure 4.15 *The geometry of multiple reflection in an isospeed shallow sea*

Next, we examine the number of surface and sea-floor bounces, for each m'th ray (of four). For the ray r'_{m1}, being the lower m'th ray above the true channel, we develop a bounce sequence by examining the left-hand part of the figure, which is: 0 (direct path, m = 1); S,B (for m = 2); S,B,S,B (for m = 3); ... . This observation we generalise, so that we may write, for the lower m'th ray above the true channel, that the overall reflection loss is μ'_{m1}, given by

$$\mu_b^{m-1}(\theta'_{m1}) \; \mu_s^{m-1}(\theta'_{m1})$$

where θ'_{m1} is the angle from the lower m'th ray to the normal to the reflecting surfaces. Proceeding in like manner for the other three m'th rays, we find for the upper m'th ray above the channel, an overall reflection coefficient, μ'_{m2}, given by

$$\mu_b^{m-1}(\theta'_{m2}) \; \mu_s^{m}(\theta'_{m2})$$

For the two rays below the channel, there will be overall reflection coefficients, μ''_{m1} and μ''_{m2}, given by

$$\mu_b^m(\theta''_{m1}) \quad \mu_s^{m-1}(\theta''_{m1})$$

and

$$\mu_b^m(\theta''_{m2}) \quad \mu_s^m(\theta''_{m2})$$

respectively. We are now in a position to express the sound pressure field at the receiver location. We write, for the source signal, that it should be (the real part of) the unit phasor $\exp(j\omega t)$. The received pressure field will then be (the real part of)

$$\exp(j\omega t) \sum_{m=1}^{\infty} \frac{\mu'_{m1} \exp(-jkr'_{m1})}{r'_{m1}} + \frac{\mu'_{m2} \exp(-jkr'_{m2})}{r'_{m2}} + \frac{\mu''_{m1} \exp(-jkr''_{m1})}{r''_{m1}} + \frac{\mu''_{m2} \exp(-jkr''_{m2})}{r''_{m2}}$$

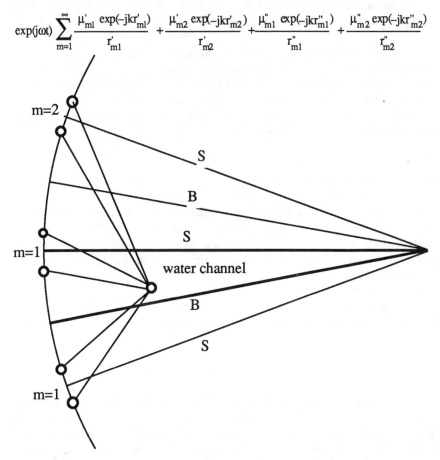

Figure 4.16 Location of source image-pairs on a circle with its centre at the intersection point between a sloping sea-floor and the sea-surface

Here we see the four m'th paths combining additively, with inverse square (power) spreading and thus inverse (first-order) pressure decrease with range. We see, also, the effect of loss caused by multiple bounces from the sea-surface and sea-bed. Finally, in the argument of the "exp" terms contained within the summation, we see the effect of transport delay between image sources and receiver. Numerically, the evaluation of this equation is straightforward. Such is the power of the modern computing workstation that the direct application of this result can yield useful insight into propagation in the shallow seas. Because of the isovelocity propagation conditions, we note that calculation of the pressure field at a point remote from the source can be achieved without recourse to conventional ray-trace methods. In this respect, the method anticipates some of the computational economies which result from the application of normal mode theory, which we study in Chapter 5.

The method discussed above may be extended to include the situation in which the sea-floor is sloping. Clearly, this circumstance is of interest in shelf-sea and continental-slope working. In this event, the image sources form themselves about a circular locus, whose centre is the (perhaps hypothetical) point of intersection of the sea-floor and sea-surface, figure 4.16. For further details, the reader is referred to [4.7].

4.6 The Lloyd Mirror Effect

One final subset of the modelling procedures discussed in the previous section is worth alluding to. If sound reflects from the surface of the sea, at such range and in such water depth, that no significant bottom reflections are encountered, then the geometry illustrated in figure 4.17 will be encountered.

If the signal transmitted is a unit sinusoid $\cos(\omega t)$, then at the receiver location, both the source and a single image in the surface will be observed. The received signal will thus be

$$\cos(\omega t - kr_1) - \cos(\omega t - kr_2)$$

where $r_2 = r_1 + c\tau$, where the negative sign between the terms indicates phase inversion on reflection from the sea-surface, and where τ is the excess propagation delay on the reflection path, by comparison with the direct path. Expanding and manipulating, we find that the mean intensity at the receiver is given by

$$I = 2r^{-2}(1 - \cos \omega \tau)$$

If next we turn our attention to the excess delay then, by inspecting figure 4.17, we see, by applying Pythagoras's Theorem to find the difference between the direct and image path ranges and simplifying by means of the Binomial Theorem, for horizontal displacements, r, significantly greater than source and receiver depths, that $\tau \approx 2z_1z_2/rc$. Consequently

$$I = 2r^{-2} (1 - \cos(4\pi z_1 z_2/r\lambda))$$

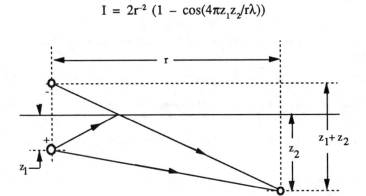

Figure 4.17 The Lloyd Mirror effect: geometry of a single sea-surface reflection

This equation shows that the channel transfer function will exhibit constructive and destructive interference which will decrease in spatial frequency with increasing horizontal displacement, as figure 4.18 illustrates. This phenomenon is known as the Lloyd Mirror effect and can sometimes be observed on sidescan sonar records, because of a sea-bed, rather than a sea-surface image interference. We shall also see, in Chapter 10, that it may prove deleterious in the operation of some sub-sea acoustic communication systems.

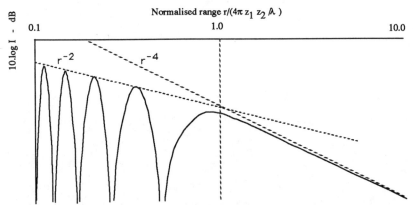

Figure 4.18 The Lloyd Mirror effect: attenuation as a function of normalised range

Note also that if r becomes very large, the argument of the cosine term will become small. The cosine term may then be expanded as a power series, of which only the first term will be significant: $\cos \xi = 1 + \xi^2/2!$ It follows that $I \propto r^{-4}$; the intensity falls as the fourth power of range.

References

[4.1] R.J. Urick, *Sound Propagation in the Sea*, Peninsula Publishing, Los Altos, Calif., 1982, pp. 3.1-3.8

[4.2] C.S. Clay and H. Medwin, *Acoustical Oceanography*, Wiley, New York, 1977

[4.3] *Physics of Sound in the Sea, Vol. 1: Transmission*. Reprinted and distributed by the Research Analysis Group, US National Research Council

[4.4] I. Tolstoy and C.S. Clay, *Ocean Acoustics*, McGraw-Hill, New York, 1966, pp. 33-36.

[4.5] C.B. Officer, *Introduction to the Theory of Sound Transmission with Application to the Ocean*, McGraw-Hill, New York, 1958

[4.6] R.J. Urick, *Sound Propagation in the Sea*, Peninsula Publishing, Los Altos, Calif., 1982, p.4.11

[4.7] J.D. Macpherson and M.J. Daintith, Practical Model of Shallow Water Acoustic Propagation, *J. Acoust. Soc. Am.*, Vol. 41, No. 4, 1967, pp. 850-854

5 Normal Mode Modelling of Sonar Propagation

co-authored by P.A. Willison

5.1 Introduction

As we saw in the previous chapter, sound propagation in shallow water leads us, via the ray tracing approach, to a model of propagation wherein there will exist a multiplicity of reflected image sources. Given iso-speed conditions, the ray paths from these sources will be straight lines. In water whose depth is moderately shallow with respect to range, there may be sufficiently few bounces for the problem of estimating the summed sound intensity developed by each ray to be computationally viable. For channels which are extremely long by comparison with water depth, the problem rapidly becomes intractable, although with the increasing power of modern scientific workstations, this difficulty is less significant than it once was.

Normal mode modelling provides an alternative to the ray trace approach. It relies upon particular solutions to the wave equation and yields an expression for the pressure field at a point remote from the sound source. We shall discuss the interpretation and development of the normal mode method in the context of an iso-speed channel over a horizontal reflecting bottom. The reader is advised that the method may be extended in a number of ways, to include propagation conditions which vary with depth and range, as well as variable bottom geometry.

We begin by establishing an understanding of the nature of normal modes in the sound channel. To emphasise an important point, imagine that a guitar string is plucked. The string will vibrate, making a half-cycle sinusoidal displacement in space, with nodal points at the bridge and nut, and a maximum amplitude of displacement at its centre. This is the fundamental mode of vibration of a stretched, thin string. A skilful guitarist can pluck the string, whilst simultaneously damping very briefly the centre of the string with the little finger of the right (plucking) hand. This will cause a node to form at the centre of the string. The string will then vibrate with a full-cycle sinusoidal spatial displacement, with nodes at bridge, nut and centre. The sound thus produced will be the first overtone, an octave up on the fundamental frequency. Second and even third overtones can be achieved by damping the string one third and one quarter of the way from the bridge.

Clearly, the stretched string may be persuaded to enter any of a number of vibrational states. These allowed states may be likened to normal modes of vibration in a waveguide in underwater acoustics. The essential condition, which dictates possible frequencies of vibration (apart that is, from distance between the bridge and nut, string density and tension) is the requirement of the existence of nodal points at the ends of the string. It is, quite simply, impossible to conceive of a vibrational state wherein the ends of a thin string are flapping up and down, and the nodal points are somewhere other than at the ends of the string. Such a situation makes no physical sense.

5.2 A Heuristic Treatment of Normal Modes in an Acoustic Waveguide

In figure 5.1, we imagine a "billiard ball" of high-pressure (the isolated dark "blob") to be incident upon the surface of the sea. Like a real billiard ball, its horizontal component of velocity will remain unchanged by the impact but its vertical component will be reversed, so that it will appear to "reflect" from the water surface. Curiously, another effect will also occur. On reflection the "billiard ball" of high-pressure will become one of low-pressure, relative to ambient (or atmospheric) pressure. This is because, being a "free surface", the pressure at the sea-surface must remain at ambient. Were the wave incident on a rigid surface which could sustain compression of the adjacent water, then the incident high-pressure "billiard ball" would reflect as a high-pressure "billiard ball". This effect occurs on reflection from the sea-floor.

In fact, the isolated "billiard ball" of high pressure has no physical parallel; fluid flow outwards from it would dissipate its pressure towards the ambient. However, a plane wavefront, such as is also shown on figure 5.1, may be likened, at least in horizontal section, to a row of "billiard balls" of high pressure. As each "ball" reaches the free surface it will reflect and pressure-invert, to form a row of low pressure "balls". That is, our high-pressure (relative to ambient) plane wavefront will reflect to become a low pressure (relative to ambient) plane wavefront. Both before and after reflection the wavefront will be travelling obliquely to the sea-surface, at the speed of sound in the sea.

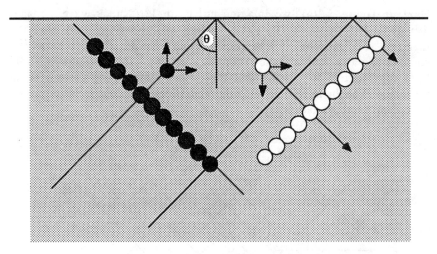

Figure 5.1 Illustrating the reflection of a section of plane wavefront from the sea-surface

In figure 5.2, we see the high-pressure incident plane wave (the darker band moving upwards towards the surface, and upon contact, developing the low-pressure reflected wave). Although the direction of travel is, in both cases, normal to the wavefront, the appearance within the frame of the picture is of an inverted "vee" running (in this example) rightwards. The apparent horizontal speed of the "vee", if the angle of incidence of the ray to the normal to the sea-surface is θ, will be $c \sin \theta$.

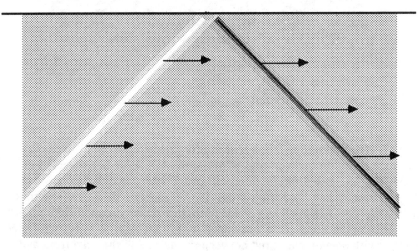

Figure 5.2 The plane wavefront, in section, moving rightwards and continuously reflecting from the sea-surface

Next, in figure 5.3(a), we imagine not a single ("shock-wave") plane-wavefront but a periodically excited plane-wave pressure field. Now the inverted "vees" must have a separation (normal to the wavefronts themselves) equal to one wavelength at the excitation frequency. In this figure, the bold lines represent high-pressure and the dashed lines low pressure, being presumed to be the maximum and minimum pressure values for a sinusoidal spatial pressure variation.

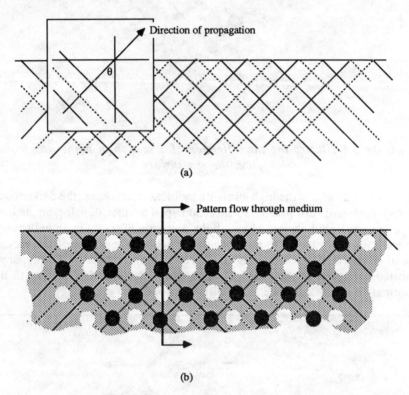

Figure 5.3 The periodic excitation of the channel, forming a rightwards moving pressure pattern

Clearly, the pressure maxima will reinforce where two bold lines cross, to produce a region of even higher pressure. A similar but reversed effect, producing increased rarefaction, will occur where two dashed lines cross. Where dashed and bold lines cross, the pressures will cancel to ambient. These effects are shown as a shaded pressure density map in figure 5.3(b). The entire pressure pattern will appear to propagate rightwards with speed $c \sin\theta$, with θ again measured between the normal to the wavefront and the normal to the sea-surface.

Now, like the guitar string which cannot have free, flapping ends, the pressure field, if it is constrained also by a rigid lower boundary, can only exist at a given frequency and angle of incidence if water depth is such as to engender the correct bottom reflection conditions so that phase reversal does not occur. Thus figure 5.4 shows the allowable depths which satisfy this constraint. Also superimposed upon this picture are sketches showing the variation of pressure amplitude with depth. This latter function will vary sinusoidally with time, at the excitation frequency.

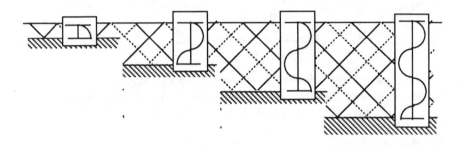

Figure 5.4 The allowable normal modes for channels of varying water depth excited by a harmonic source of fixed frequency. In the inset figures, pressure is plotted horizontally, as a function of depth

Of course, we cannot choose channel depth, h, to suit our own convenience, so in practice we find that an appropriate normal mode of vibration can develop, at a given drive frequency, only at a particular angle of incidence. Reference to figure 5.5 shows that this angle, θ, is given by

$$\theta = \cos^{-1}(\lambda/4h)$$

for the first mode and, by extension, is given for higher modes by the expression

$$\theta_n = \cos^{-1}((\lambda/2h)(n - 1/2)); \quad n = 1, 2, 3 \ldots$$

This equation, which is known as the characteristic equation, is a variant of the Bragg equation which governs the interference pattern of a diffraction grating. Weston [5.1] based a series of ripple tank experiments on this relation. We note, once more, that once the channel depth and the transmission frequency are fixed, only certain angles of incidence are allowed for propagation along the waveguide.

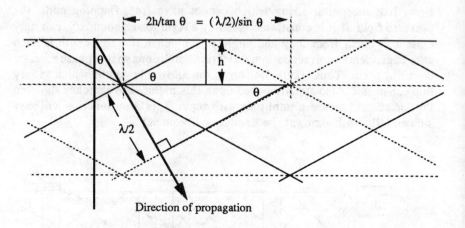

Figure 5.5 Illustrating the derivation of the inter-relationship between angle of incidence, θ, wavelength, λ and water depth, h, for the first normal mode

We also note that, for each mode, because of the orientation of the wavefronts to the channel boundaries, the speed of pressure interference pattern propagation down the channel, which is known as the group velocity, will be given for each mode by the expression

$$u_n = c \sin \theta_n$$

If θ_n tends to 90°, the plane wavefronts approach vertical alignment and pass down the channel at a speed close to the speed of sound in free space. This condition is approached only as the excitation frequency becomes very high.

When the excitation frequency for the first mode falls to such a value that the channel depth is spanned by one quarter wavelength, we encounter a situation where the plane wavefronts form only standing waves in the vertical, reflecting continuously backwards and forwards between the sea-surface and sea-floor. Then the group velocity falls to zero, $\theta_n \to 0°$, and no forward propagation can occur.

More generally, this effect will occur when the excitation wavelength for the n'th mode has fallen to a value satisfied by the relationship

$$h = (\lambda_{cn}/2)(n - 1/2);$$

that is, when the frequency $f_{cn} = c/\lambda_{cn}$ is given by

$$f_{cn} = (c/2h)(n - 1/2)$$

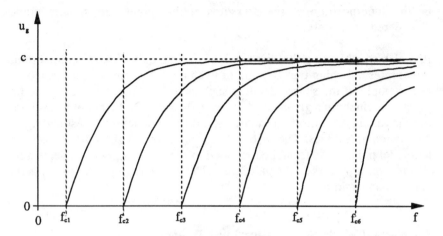

Figure 5.6 The variability of group velocity with frequency

This frequency is referred to as the cutoff frequency of the n'th mode. For frequencies above the cutoff, propagation can occur and will take place at an incidence angle to the surface normal of θ_n and group velocity u_n. Since the geometry of figure 5.5 allows us to express θ_n in terms of the ratio of wavelengths at excitation frequency and cutoff frequency, λ and λ_{cn} respectively, as

$$\sin^2 \theta_n = \frac{\lambda_{cn}^2 - \lambda^2}{\lambda_{cn}^2}$$

we may derive a new expression for group velocity in terms of mode excitation and cutoff frequencies which is

$$u_n = c[1 - (f_{cn}/f)^2]^{-1/2}$$

This expression is shown plotted in figure 5.6. If the excitation frequency were, say, half-way between the fourth and fifth mode cutoff frequencies, we should expect to experience the lowest four modes propagating in the duct. Whilst this would be true if the sea-surface and sea-floor were perfect reflectors, practical conditions will tend to reduce the number of modes which can propagate effectively.

We thus anticipate a situation in which our harmonic source will excite a certain number of modes within the acoustic waveguide. These several modes will simultaneously co-exist. They will propagate, each at its own group velocity, thus establishing a pressure at some remote point which will be the superposition of the pressures at that point due to each mode considered separately.

It is also worth noting that the depth of the source will have an important effect in establishing the strength, or even the existence, of a given mode. For example, if the source depth is such as to locate the source at a nodal point, for which for a given mode $p(z) = 0$, then that mode cannot be excited. For maximum excitation, the source should be at a depth corresponding to one of the anti-nodes. The inset curves in figure 5.4 show the nodal and antinodal points. The former correspond to zero-crossings on the depth axis of the pressure versus depth curves. The latter correspond to pressure maxima and minima.

5.3 Normal Mode Solution for Long Ranges

The reason for developing the normal mode solution was that the method of images involves, in principle at least, an infinite summation of rays. Because of this, application of the method, particularly in an inhomogeneous medium and at long ranges, can be computationally cumbersome. We have yet to show that the mode solution improves on the image approach. Indeed, we have deduced that, to describe accurately the pressure field at a distant point, we require an infinite sum of modes so that, at first sight, matters might not be thought to have improved. We will now show that, assuming that we *are* interested in long range propagation, the pressure at a point can be found by the summation of a finite number of normal modes.

The model to be employed retains, for simplicity only, the homogeneous water layer and free surface condition, as in section 5.2, but now employs a "non-rigid" bottom. If a wave is incident on the bottom boundary with an angle less than some critical angle, the wave is partially reflected back into the water layer and partially transmitted into the lower layer. The critical angle is given by

$$\theta_c = \sin^{-1}(c/c_{sed})$$

where c and c_{sed} are the sound speeds in the water and in the lower, sedimentary sea-floor respectively. We assume that, for angles of incidence greater than the critical angle the wave undergoes total internal reflection and no energy is lost to the lower layer. This is an idealisation of the reflection characteristic for the "fast bottom" which is illustrated in figure 1.6. To avoid the problem of energy being transmitted back into the surface layer we assume that the lower layer is infinitely thick.

The modes for which (because of angle of incidence) energy is partially transmitted into the lower layer will be quickly attenuated as a result of this energy leakage, and thus will have no significant bearing on the acoustic pressure field at long ranges. To calculate the pressure at long range we need only be interested in the modes that represent the guided or trapped waves.

To determine the highest mode that we need to consider, we rearrange the characteristic equation to give

$$n = (2hf/c) \cos \theta_c + 1/2$$

where θ_c is the critical angle measured from the normal. The highest mode of interest, m, is then given by rounding n to the next lower integer value. To calculate the pressure at a distant point we therefore sum from 1 to m, assuming the transmit frequency is higher than the cutoff frequency of the first mode. The attraction of the Normal Mode method is now readily apparent. The solution of the long range propagation problem is simply given by summing a finite, often small, number of modes.

Let us consider an example. Assume a channel depth of 7.5 m. The channel cutoff will then be 50 Hz for the first mode, 150 Hz for the second, 250 Hz for the third, and so on. This calculation applies strictly to the channel with a rigid bottom. Here, as figure 5.7(a) illustrates, we see the pressure fluctuation amplitude (about ambient) decrease immediately to zero, below the perfect reflecting rigid sea-floor. The modal patterns of pressure amplitude with depth are as we have been led to expect from our previous discussions.

In contrast, if the sea-floor boundary is not perfectly reflecting, then we should expect sound to penetrate and for there to be a reasonably rapid decrease of amplitude with increasing depth into the sediment, because of the high attenuation within the sediment. This situation is shown in figure 5.7(b). The pressure amplitude will be the same in the sea just above the sediment as it will be just within the sea-floor. This reflects a requirement for continuity of pressure at this interface. If there is still some sensible difference between the density of the sea-water and the density of the sediment then, although an "ideal" waveguide will not exist, and the sub-bottom half-space will undoubtedly influence the modal cutoff frequencies, there will yet remain at least a loose correspondence between the cutoff frequencies for the real and for the ideal channel. For a fuller treatment of this topic, the reader is referred to any of the advanced texts which specifically treat sound propagation in the sea, of which several are listed at the end of this chapter [5.4 – 5.14].

For our tutorial purposes, we accept that the mode cutoff frequencies, as indicated by considering behaviour with a rigid bottom, are at least adequate in predicting the approximate maximum number of modes which could propagate, at a given excitation frequency.

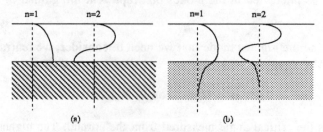

Figure 5.7 Showing pressure variation with depth for the first two modal states: (a) for a rigid sea-floor and (b) for a sea-floor, such as an unconsolidated sediment, which is not perfectly reflecting and into which sound may penetrate

For example, suppose we attempted to excite the channel with a 2000 Hz source. Then we should expect, with a perfectly rigid bottom, that some twenty modes would propagate, albeit with decreasing sound speed as mode number increased. We now need to estimate how many of the higher of those modes might safely be ignored, by virtue of a systematic loss of energy into the sediment, for incidence angles greater than critical.

Suppose for example, that sound speed in the water layer is taken as 1500 m/s and in the sediment layer is arbitrarily chosen as 1600 m/s. This is the condition of the so-called "fast bottom" which might well correspond to our unconsolidated sea-bed composed of fine sand. We can now calculate the critical angle, θ_c as

$$\sin^{-1}(1500/1600) = 70°; \cos\theta_c = 0.34$$

The number of modes of interest is then

m=next integer value below $\{((2 \times 7.5 \times 2000 \times 0.34)/1500) + 1/2\} = 7$

We thus need only calculate the pressure field by summing the effects of the first seven modes, rather than all of the twenty which could in principle propagate with greater than zero group velocity, since energy leakage into the sea-bed will mean that the thirteen highest modes will not propagate to long ranges at 2000 Hz.

The advantage of the normal mode solution is now obvious, with the reduction of an infinite summation to that of a relatively few terms. It should be stressed that this is only an approximate solution and is only valid for ranges in excess of a few water depths. At short ranges the influence of the rapidly attenuated modes must be included. Additionally, we should include the effect of a non-rigid bottom more comprehensively than is possible here. These situations make for a far more complicated analysis and the necessary extensions are reviewed briefly in section 5.6.

Another factor to be borne in mind when considering propagation to long ranges is that the channel will behave in a markedly dispersive manner, because of the variability of group delay with frequency. Sound from an explosive source, such as might be used in marine seismic survey, will contain a rich distribution of spectral components in the audio range spanning, in the main, the 200 Hz to 2 kHz regime used as an example in the previous paragraph.

Heard close-to, the detonation will sound like a deeply pitched "crunch". When detected remotely in water which, near the coast, may have shallowed over many hundreds of metres to a depth of only a few metres, the detonation will sound remarkably like a down-chirping sinusoidal sonar pulse of a second or so duration, with a start-frequency of about 2 kHz and an end-frequency in the low hundreds of Hz. The reason why this is so is because, at the higher frequencies, the group velocity of most of the (9 or so, by our previous calculation) normal modes will be clustered towards the (asymptotic) free-field sound speed, c. The lower frequencies will sustain fewer and fewer modes and will exhibit group delay values significantly less than the free-field value. The higher frequency components will thus arrive first and with lower loss, producing the down-chirp effect.

5.4 Normal Modes as Interfering Plane Waves

To give a more mathematical interpretation of the propagation of sound as normal modes, we shall show that the modes can be generated by the summation of up-going and down-going plane waves. We will then be able to show the connection with the normal modes and the method of images discussed earlier. We can represent an up-going and a down-going plane wave by the first and second terms, respectively, in the expression

$$p = A_1 \exp(j(\omega t - \kappa r - \gamma z)) + A_2 \exp(j(\omega t - \kappa r + \gamma z))$$

where κ is the horizontal component and γ the vertical component of wave number $k = \omega/c$, measured along the propagation direction θ, such that $k^2 = \kappa^2 + \gamma^2$. At the surface, when $z = 0$, the contributions from the plane waves must sum to zero:

$$A_1 \exp(j(\omega t - \kappa r)) + A_2 \exp(j(\omega t - \kappa r)) = 0$$

and it follows that $A_1 = -A_2 = A$. At the bottom, when $z = h$, the pressure must be allowed to be a maximum, so that $dp/dz = 0$.

Manipulating, we find that

$$dp/dz = 2jA\gamma \exp(j(\omega t - \kappa r)) \cos\gamma h = 0$$

which can only occur when $\gamma h = \pi/2, 3\pi/2, 5\pi/2 \ldots$ or

$$\gamma = \frac{\pi}{h}(n - 1/2); \quad n = 1, 2, 3\ldots$$

Now, since

$$\gamma = k \cos(\theta_n)$$

we find, again, that

$$\theta_n = \cos^{-1}((\lambda/2h)(n - 1/2)); \quad n = 1, 2, 3 \ldots$$

As the frequency of the excitation is decreased, the angle of incidence (between the surface or sea-floor and the normal to the wavefronts) becomes larger. The frequency at which the angle is 90° defines, again, the mode cutoff. When in this condition, we see once more that we have a purely standing wave in the z direction (upwards and downwards) with no forward propagation component. Consequently, it is obvious that for propagation we require the transmission frequency to be greater than the cutoff frequency of the first mode.

5.5 The Normal Mode Solution Formalised

We continue by retaining our assumption of the iso-speed channel model with an ideal pressure release surface and a rigid bottom. At locations which do not include any sources, the acoustic pressure field $p(x,y,z,t)$ in a water layer can be shown [5.2], to obey the wave equation

$$\nabla^2 p = \frac{1}{c^2} \frac{\partial^2 p}{\partial t^2}$$

Normal Mode Modelling of Sonar Propagation

We shall show that the acoustic pressure field can be represented by a linear superposition of travelling normal modes. In a physical sense, as we have seen, the normal modes describe the way in which the fluid is vibrating, akin to the vibrations of a taut string when plucked. The wave equation is most easily solved by the method of separation of variables. For our model we have the 2-dimensional wave equation

$$\frac{\partial^2 p}{\partial x^2} + \frac{\partial^2 p}{\partial z^2} = \frac{1}{c^2}\frac{\partial^2 p}{\partial t^2}$$

The pressure field extending from a source between plane, parallel surfaces will exhibit circular symmetry in the horizontal. This suggests that it will be mathematically more convenient to solve the equation in cylindrical co-ordinates giving

$$\frac{\partial^2 p}{\partial r^2} + \frac{1}{r}\frac{\partial p}{\partial r} + \frac{\partial^2 p}{\partial z^2} = \frac{1}{c^2}\frac{\partial^2 p}{\partial t^2}$$

To simplify the solution we assume that $p(r,z,t)$ has a time dependence of the form $\exp(jwt)$. This allows us to re-write the wave equation in cylindrical co-ordinates as

$$\frac{\partial^2 p}{\partial r^2} + \frac{1}{r}\frac{\partial p}{\partial r} + \frac{\partial^2 p}{\partial z^2} + k^2 p = 0$$

where $k = \omega/c$. This result is known as the time-independent Helmholtz equation. Using the method of separation of variables, we assume a solution of the form

$$p(r,z) = R(r)Z(z)$$

Substitution gives

$$ZR'' + \frac{1}{r}ZR' + Z''R + k^2 RZ = 0$$

Dividing by ZR and re-arranging, we find that

$$\frac{R''}{R} + \frac{R'}{R} = -\frac{Z''}{Z} - k^2$$

The left-hand side of this equation is dependent only upon the range, r, whilst the right is dependent only upon the depth, z. This can only be the case when both sides are equal to a constant, referred to as the "separation constant". We write the separation constant as $-\lambda^2$ and thus obtain two ordinary linear differential equations

$$\frac{d^2Z}{dz^2} + Z\left(\frac{\omega^2}{c^2} - \lambda^2\right) = 0 \qquad \frac{d^2R}{dr^2} + \frac{1}{r}\frac{d^2R}{dr^2} + \lambda^2 R = 0$$

The general solution to the first of these equations is

$$Z(z) = A\sin(\psi z) + B\cos(\psi z)$$

where

$$\psi = \left[\frac{\omega^2}{c^2} - \lambda^2\right]^{1/2}$$

We know that at $z = 0$ we must have $Z(z) = 0$, therefore $B = 0$. The second boundary condition requires the pressure to be a maximum at $z = h$ and therefore

$$|\sin(\psi h)| = 1$$

This requires that $\psi h = \pi/2, 3\pi/2, 5\pi/2, \ldots$ or that $\psi n = (\pi/h)(2n - 1)$; $n = 1, 2, 3 \ldots$ ψn is termed an eigenvalue and the solution

$$Z_n(z) = A_n \sin(\psi_n z)$$

is an eigenfunction of the problem. We see that there is an infinity of eigenvalues and thus infinitely many eigenfunctions as solutions to the boundary value problem. By solving this equation as a Sturm–Liouville problem, see [5.3], we see that the eigenfunctions so obtained have the property that they form an orthogonal set with respect to some weight function, that is

$$\rho \int_0^h Z_m(z) Z_n(z) \, dz = \delta_{mn}$$

where d is the Kronecker delta function. Thus any single eigenfunction or any linear superposition of eigenfunctions forms a solution to the wave equation. The eigenfunctions are known as the normal modes of propagation in our perfect waveguide model. The range-dependent differential equation derived above is a form of Bessel's equation, the method of solution of which will depend upon the accuracy and range involved in a particular problem.

Exploiting the orthogonality of the eigenfunctions, we may expand the acoustic field as a sum of the normal modes to yield the pressure field equation. A typical form establishing a particular solution to the wave equation which defines pressure as a function of time, horizontal range and depth, and vertical sound velocity profile is

$$p(r,z,t) = A \exp(i(\omega t + \pi/4)) \Sigma Z_n(z_0)Z_n(z)\exp(-i\kappa_n r)/v_n(2\pi\kappa_n r)$$

where A is an amplitude factor, describing source level, the function Z_n defines the nature of allowed vertical standing waves and k_n is the horizontal component of wave number, k. The vertical component of wave number, γ_n, must be such as to ensure that a pressure node exists at the surface and, if we assume a perfectly rigid reflecting bottom, a pressure anti-node at the sea-bed. Note that the eigenfunctions essentially plot the depth dependence of pressure. For given water depth and transmission frequency, only a finite number of modes may exist. The sum of the allowed modes, as determined by the defining equation given above, establishes the pressure field. Time dependence, for a harmonic source, is determined by the leading "exp" function. Range dependence is built into the denominator function, and the contained "exp" function.

Local waveshape in the vicinity (r,z) may be obtained for non-harmonic sources, by Fourier transforming p(r,z,t) to obtain an effective channel transfer function P(r,z,f), multiplying by the source spectrum and re-transforming. Such manipulation requires obvious care, considerable skill in numerical analysis and computing machinery of some power. The method is of importance, because the channel, at low frequencies and shallow water depths, is a "dispersive" waveguide. That is, as with light split by a glass prism into its spectral components, sounds of differing frequencies travel with different speeds in the guide, a phenomenon not observable in free-field propagation.

5.6 Normal Mode Solution for All Ranges

In section 5.3, we saw that the Normal Mode approach exploited the fact that only those modes that represent the trapped energy are of any great significance at long ranges and therefore the modes that represent the energy which leaks out of the channel could be ignored. For an exact solution, applicable to all ranges, we must include those modes that were ignored. This is a more difficult problem, but one which has received a good deal of interest over the last forty years. Pekeris [5.4] is generally regarded as having pioneered the waveguide approach to the solution of acoustic propagation in the shallow water channel.

We have seen that the eigenvalues for the perfect waveguide are quantised. The associated modes are called discrete modes. If we allow water depth, h, to tend to infinity, then the difference between successive eigenvalues will tend to zero. The eigenvalues will form a continuous set and the eigenfunctions, the normal modes, are called continuous modes. For an in-depth discussion of continuous and discrete modes the reader is referred to Tolstoy and Clay [5.5]. It is generally accepted that at short ranges, where continuous modes must be taken into account, the Ray Tracing approach is more favourable.

5.7 The Horizontally Stratified Channel

We now wish to consider the solution of the more general problem in which the sound speed and density are allowed to vary with depth. This problem is usually solved by extending the model introduced in section 5.3 to allow for many horizontal layers each with its own constant sound speed and density, or density as a known function of depth. As the width of the stratifications is reduced the model tends to that of a continuously horizontally varying medium.

The shallow water channel that is of interest to us may not contain appreciable differences in sound speed and density in the water layer. It is, however, possible that the parameters of a number of the sub-bottom layers may be known and can then be easily included in the model. Waveguide propagation will only take place when there exists a layer with a sound speed minimum relative to the other layers.
Early studies by Pekeris [5.4] and by Officer [5.6] dealt with a maximum of three layers. Tolstoy extended the theory to many layers using physical and geometric reasoning [5.7, 5.8] to generate the characteristic equation for the normal modes. Other authors, such as Budden [5.9], Brekhovskikh and Lysanov [5.10], Brekhovskikh [5.11] and Clay and Medwin [5.12] have employed similar approaches. More recent mathematical treatments, such as those given by Stickler [5.13] and by Boyles [5.14], all solve the wave equation with varying degrees of accuracy.

References

[5.1] D.E. Weston, A Moire Fringe Analog of Sound Propagation in Shallow Water, *J. Acoust. Soc. Am.*, Vol. 32, No. 6, 1960, pp. 647-654

[5.2] L.E. Kinsler, A.R. Frey, A.B. Coppens and J.V. Sanders, *Fundamentals of Acoustics*, Wiley, 3rd edition, New York, 1982

[5.3] E. Kreyszig, *Advanced Engineering Mathematics*, Wiley, 5th edition, New York, 1983

[5.4] C.L. Pekeris, Theory of Propagation of Explosive Sound in Shallow Water, in *Propagation of Sound in the Oceans*, Geol. Soc. Am. Memoir 27, 1948

[5.5] I. Tolstoy and C.S. Clay, *Ocean Acoustics*, McGraw-Hill, New York, 1966, pp. 33-36

[5.6] C.B. Officer, *Introduction to the Theory of Sound Transmission with Application to the Ocean*, McGraw-Hill, New York, 1958

[5.7] I. Tolstoy, Shallow Water Test of the Theory of Layered Wave Guides, *J. Acoust. Soc. Am.*, Vol. 30, No. 4, 1958, pp. 348-361

[5.8] I. Tolstoy, Note on the Propagation of Normal Modes in Inhomogeneous Media, *J. Acoust. Soc. Am.*, Vol. 27, No. 2, 1955, pp. 274-277

[5.9] K.G. Budden, *The Wave-Guide Mode Theory of Wave Propagation*, Logos, London, 1961

[5.10] L. Brekhovskikh and Y. Lysanov, *Fundamentals of Ocean Acoustics*, Springer-Verlag, Berlin, 1982

[5.11] L.M. Brekhovskikh, *Waves In Layered Media* (R.T. Beyer, transl.), 2nd edition, Academic Press, London, 1980

[5.12] C.S. Clay and H. Medwin, *Acoustical Oceanography*, Wiley, New York, 1977

[5.13] D.C. Stickler, Normal-mode Program with Both the Discrete and Branch Line Contributions, *J. Acoust. Soc. Am.*, Vol. 57, No. 4, 1975, pp. 856-861

[5.14] C.A. Boyles, *Acoustic Waveguides*, Wiley-Interscience, New York, 1984

6 Noise and Reverberation

6.1 Introduction

The study of acoustic noise and the reverberation of acoustic signals is of importance because one or other of these phenomena will set the limit to sonar system performance. Whilst it is true that both sources of corruption can co-exist, it is most common to find that one of the two will predominate. Notice that in a noise-limited sonar, increasing the signal power will improve the signal-to-noise ratio and hence the system performance. The same need not be true for a reverberation limited sonar, since the reverberation is, itself, directly a function of the output signal level.

Noise in the sea may derive from many sources: seismic events, wave action, shipping, thermal agitation, rainfall, sounds made by marine animals and so on. The ambient noise field is considered isotropic, in the sense that, wherever a sensing hydrophone is placed in the sea, the observed intensity of the noise will remain the same – subject of course to a similarity of relevant environmental conditions. This does not mean, however, that the noise is not directional. It is easily possible to envisage a directional hydrophone sensing a higher angular intensity density of noise, when viewing along the axis of a sound channel such as would form at the base of the main thermocline (see section 6.3, for example), than it would when looking vertically upwards or downwards.

Many noise sources would present a continuous spectrum together with gaussian statistics. This would certainly be true of thermal noise at high frequencies but would also hold for wave and rain-induced noise. By contrast, shipping noise would present a mixed spectrum with both continuous and discrete components, the former deriving largely from the bow-wave and propeller cavitation, the latter from machinery noise. Whilst the bow-wave might well present gaussian and stationary statistics, it is quite likely that propeller cavitation, although also a noise process, would exhibit short-term non-stationary behaviour producing a measurable pulsing of acoustic intensity at a frequency proportional to the blade-rate.

6.2 Deep Sea Ambient Noise Level [6.1]

In using the term "ambient", we discount such local effects as the "self-noise" of a moving vessel which might be carrying a sonar. Self-noise is caused by the passage of the vessel through the water and by vibrations induced by its machinery and propellers. Note that we shall not need to correct ambient noise level for range in sonar calculations because it is an all-pervading quantity, much as heat in a greenhouse, though having the sun as a "localised" source, is all-pervading.

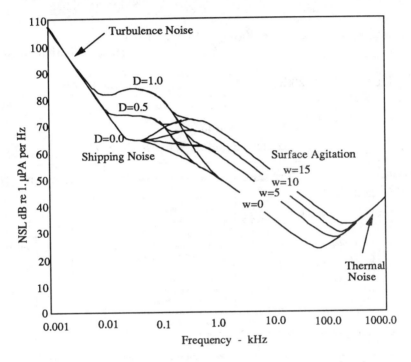

Figure 6.1. Deep-water ambient noise spectrum level

We commence by examining, figure 6.1, the gross spectral behaviour of acoustic noise observed in the open sea. The lowest frequency decade, between 1 Hz and 10 Hz, is dominated by noise originating from oceanic turbulence. Over the next frequency decade, shipping is the dominant cause. Noise in the next three decades, between 100 Hz and 100 kHz, is primarily caused by surface agitation. Finally, at frequencies in excess of about 100 kHz, it is thermal noise, originating in molecular motion in the sea which is the principal effect. The overall noise spectrum level, NSL, or the intensity attributable to a spectrum measurement bandwidth of 1 Hz, is the power sum of the noise spectrum levels NSL1 to NSL_4 attributable to these four predominant sources of noise. We may empirically calculate the individual noise spectrum levels as

$$NSL1 = 17 - 30(\log f) \qquad \text{Turbulence Noise}$$

$$NSL_2 = 40 + 20(D - 0.5) + 26(\log f) - 60(\log (f + 0.03))$$
$$\text{Shipping Noise}$$

$$NSL_3 = 50 + 7.5\, w^{1/2} + 20(\log f) - 40(\log(f + 0.4))$$
$$\text{Surface Agitation Noise}$$

$$NSL4 = -15 + 20(\log f) \qquad \text{Thermal Noise}$$

Table 6.1
Nautical Beaufort Wind Scale

Beaufort Number	Wind Name	Wind Speed knots	Wind Speed m/s	Description of Sea Surface	Sea state	Mean wave ht ft	Mean wave ht m
0	calm	<1	<0.5	sea mirror-like	0	0	0
1	light air	1-3	0.5-2	scale-like ripples, no foam-crests	0	0	0
2	light breeze	4-6	2-3	small wavelets, crests glassy but not breaking	1	0-1	0-0.3
3	gentle breeze	7-10	3-5	large wavelets, crests begin to break	2	1-2	0.3-0.6
4	moderate breeze	11-16	5-8	small waves, fairly frequent white horses	3	2-4	0.6-1.2
5	fresh breeze	17-21	8-11	moderate waves, many white horses	4	4-8	1.2-2.4
6	strong breeze	22-27	11-13	large waves, white foam crests, some spray	5	8-13	2.4-4
7	moderate gale	28-33	14-16	sea heaping; foam begins to be blown in streaks	6	13-20	4-6
8	fresh gale	34-40	17-20	moderately high waves, marked foam streaking	6	13-20	4-6
9	strong gale	41-47	21-24	high waves, rolling sea, spray reduces visibility	6	13-20	4-6
10	whole gale	48-55	24-27	v. high waves, overhanging crests, sea white	7	20-30	6-9
11	storm	56-65	28-33	except. high waves, ships lost to visibility	8	30-45	9-14
12+	hurricane	65+	34+	air filled with foam; visibility v. poor	9	45+	14+

where f is frequency in kHz, D is shipping density on a scale 0 (very light) to 1 (heavy) and w is windspeed, in metres per second (1 m s^{-1} ≈ 2 knots

≈ 2 mph). Table 6.1 summarises the nautical Beaufort scale of wind. We then calculate the overall noise spectrum level as

$$NSL = 10 \log \{ 10^{NSL_1/10} + 10^{NSL_2/10} + 10^{NSL_3/10} + 10^{NSL_4/10} \}$$

Finally, to allow for an actual transmission bandwidth of B Hz, we calculate the noise level as

$$NL = NSL + 10 \log B$$

6.3 The Variability of Ambient Noise with Time

In contemplating the general utility of the curves presented in figure 6.1, we are drawn immediately to question the stationarity of the various causative processes. Clearly, over some identifiable time-scale, all the underlying causative processes, except for thermal noise, must be considered exceedingly variable. It is, in general, considered that noise due to shipping exhibits the shortest-term variability. Wind-induced noise, by contrast, exhibits statistics which vary relatively slowly because of the inertia imparted by the time required to build up, or dissipate, a "fully-arisen" sea. The following broad generalisations are worthy of note.

1. The ambient noise spectrum has not, thus far, been observed to yield evidence of seasonal variability.

2. Short-term variability of wind-induced noise has been observed to follow the loose pattern indicated by figure 6.2. Typically the noise samples used to estimate standard deviation, as depicted here, were of the order of 100 seconds' duration and were taken at intervals of one hour [6.2].

3. In shallow water, ambient noise exhibits wider, more rapid variability, particularly in the vicinity of harbours and shipping lanes. This variability also exhibits longer period, diurnal fluctuations because of the reduction in traffic during the night.

4. Coastal locations where breaking surf is present also exhibit a significant increase of, perhaps, some 10 dB over deeper water predictions. In the main, however, the general form of the spectrum noise level curves mirror those shown in figure 6.1.

5. In all these various aspects, available data concerning variability is poorly co-ordinated and often inadequately documented in so far as precise description of experimental method is concerned. Appropriate statistical descriptors for both magnitude variability and duration remain yet to be properly developed.

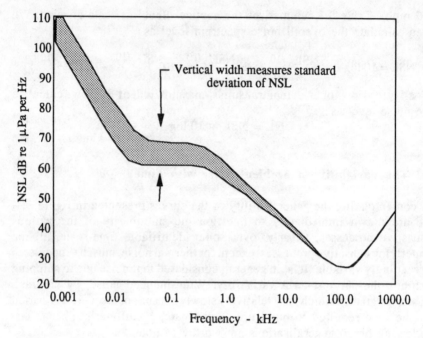

Figure 6.2. The gross short-term temporal variability of noise spectrum level

6.4 The Variability of Ambient Noise Level with Depth

In general, ambient noise level decreases with increasing depth. This is because the noise is generated at the sea-surface by wind or shipping, and because, as depth increases, so also does range-dependent attenuation caused by sound absorption. However at frequencies in the band between 10 and 100 Hz, where shipping noise may predominate, there is in deep ocean water only a small decrease with depth, because of duct-like propagation within the deep ocean sound channel.

The deep ocean sound channel has its origin in the nature of the deep ocean sound speed profile, which we examined in section 4.3. Figure 6.3 shows a simplified sound speed profile, together with a ray launched at that depth, z_1, at which the sound speed is a minimum. The launch angle, θ_1, has been so chosen that the ray just grazes the surface. Were the angle (upwards, towards the vertical) only marginally smaller, we should expect reflection, rather than grazing, at the ocean-surface, and thus (relatively) high loss. Note that the characteristic ray parameter, **a**, is given by $\sin\theta_1/c(z_1)$. The same ray, passing below the depth z_1 will curve circularly and become horizontal at a new depth z_2, for which – because of Snell's law – $c(z_2)$ must equal $c(0)$. The ray launched at an angle θ_1 downwards will also behave in similar manner, by virtue of the geometry of the ray construction. All rays between these two extremes of angle will also snake forward along the deep sound channel axis at depth z_1 and all will exhibit only cylindrical spreading loss.

The previous paragraph describes the way in which sound may propagate along the deep sound channel if launched on the axis, at depth z_1. However, surface generated noise can also pass into the deep sound channel and become trapped there, propagating onwards with low loss. Only rays so angled that they pass below the turning depth z_2 will return to intercept the ocean surface in a reflection, thus engendering loss. It is of course possible for rays to become incident reflectively on the ocean floor where, again, loss will occur. Because of this, it is observed that at depths down to z_2, which is referred to as the critical depth, negligible decrease in spectrum noise level will occur. Below this depth a more rapid decrease will take place, as figure 6.3 also illustrates.

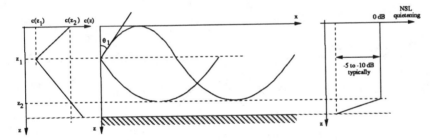

Figure 6.3. Sound trapping in the deep sound channel and the quietening of ambient noise below the critical depth, z_2

6.5 The Angular Distribution of the Ambient Noise Field

The sea-surface may be considered for most purposes to be the originating region for ambient noise over the normal operating frequency regime of the vast majority of sonar equipment. The sea-surface is frequently thought of as consisting of densely packed dipole random sources, for each of which will exist the characteristic dipole (power) radiation pattern

$$I(\theta) = I_0 \sin^2\theta$$

where I_0 is the intensity encountered in a downward vertical direction, figure 6.4. Dipole radiation is considered more fully in section 8.2.

To determine the angular distribution of intensity in a deep ocean, we consider the geometry illustrated in figure 6.5. In this figure, it is assumed that no horizontal directionality will be encountered, so we seek to sum contributions from circular annuli at the hydrophone location. We also assume that, because the ocean is deep, reflections from the ocean floor may be neglected. This is because attenuation will greatly reduce the noise contribution induced by ocean floor reflection, by comparison with that emanating from the ocean surface, at least for a shallowly immersed hydrophone. The intensity at the hydrophone will then be

$$dI = I(\theta) \, 2\pi r l^{-2} \, dr$$

where r is the radius of the annulus and l the slant range to the hydrophone. The equation thus expressed includes both the radiating area and the spreading loss along the transmission path.

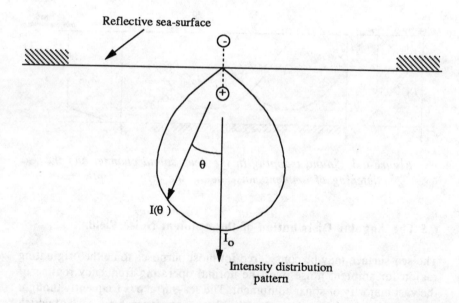

Figure 6.4. The angular intensity of a dipole source in the sea-surface

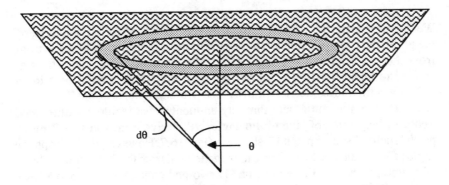

Figure 6.5. View from below of a circular annulus in the sea-surface, acting as a distributed noise source as sensed by a directional hydrophone with an upwards viewing angle θ

The annular radius and slant range, and hence the incremental intensity dI, may now be expressed in terms of hydrophone viewing angle θ and hydrophone depth z_h. We then find that

$$dI = 2\pi I_0 \cos^2\theta \tan\theta \, d\theta$$

We may recast this result to provide the intensity per unit solid angle ψ by writing

$$dI/d\psi = dI/(2\pi \sin\theta \, d\theta) = I_0 \cos\theta$$

This remarkably simple result tells us that the intensity is greatest when the hydrophone views the sea-surface directly from below. It further tells us that viewed horizontally, we expect to encounter no ambient noise contribution. Because the sea-floor was presumed infinitely far removed from the surface and the hydrophone, no noise will be received from below the vertical. Finally, we note that neither the pattern of directional characteristic of the noise, nor its magnitude, varies with depth. We thus write, for the noise intensity per unit solid angle

$$N(\theta) = I_0 \cos\theta \qquad 0 \leq q \leq \pi/2$$
$$= 0 \qquad \pi/2 < \theta \leq \pi$$

This curve is illustrated in figure 6.6 and has been found to accord reasonably well, at frequencies in excess of several hundreds of Hertz, with experimentally acquired results.

We turn next to the question of vertical directionality in s*hallow* water. Here, the problem is complicated by the possibility of a multiplicity of reflections from the sea-floor and sea-surface. For the moment, we assume the sea-surface to be an ideal reflector but the sea-floor we assume to be a lossy reflector with a loss coefficient $\mu(\theta)$ (see, for example, sections 1.10, 1.11 and 2.7). We may anticipate that any ambient noise incident within, say, a cone of inspection of given solid angle, will derive from a surface "source patch" such as "source patch 1", shown in figure 6.7. However, the hydrophone angle of acceptance will also sense noise generated from the further multiplicity of source patches 2, 3 ... and so on. The folded cone of acceptance allows us to write the summed intensity of all such source patches as

$$N(\theta) = I_0 \cos\theta(1 + \mu(\theta) + \mu^2(\theta) + ...)$$

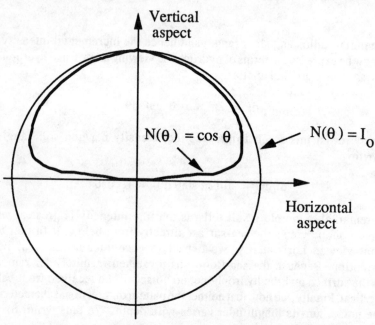

Figure 6.6. *The vertical directivity of ambient noise in a deep ocean, where ocean-floor reflections are greatly attenuated by the relatively large distance between the ocean surface and floor by comparison with the distance between the surface and hydrophone*

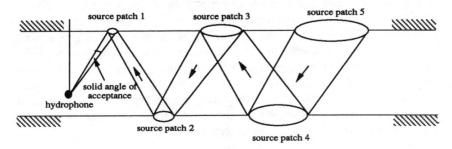

Figure 6.7. *The folded cone of inspection for a directional hydrophone looking upwards towards the sea-surface in shallow water, where multiply reflected surface noise patches contribute to the overall acoustic intensity at the receiver location*

which may be re-written as

$$N(\theta) = I_0 \cos\theta \{1 - \mu(\theta)\}^{-1}$$

Strictly, this result holds for upward-looking cones of inspection. If the cone is directed downwards, the same calculation applies, but with an extra bottom bounce, so that the computation of vertical noise directivity may be re-written in the more complete form, as

$$N(\theta) = I_0 \cos\theta \{1 - \mu(\theta)\}^{-1}; \qquad 0 \leq \theta \leq \pi/2$$

$$N(\theta) = I_0 \cos\theta\, \mu(\theta) \{1 - \mu(\theta)\}^{-1}; \quad \pi/2 < \theta \leq \pi$$

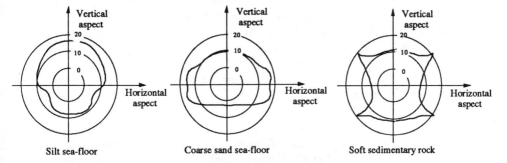

Figure 6.8. *Vertical directionality of ambient noise in a shallow sea, calculated in decibels relative to dipole field source vertical axis intensity I_o*

Again we note that the vertical directionality is independent of depth. Clearly, calculation of N(θ) depends upon the model chosen for sea-floor reflectivity. Chapman [6.3] has identified six typical sea-bed types and computed both bottom reflection loss and relative noise intensity in decibels, as a function of the direction angle θ for the cone of inspection. His results show considerable variability in directionality with sea-bed type. Figure 6.8 summarises his conclusions for three bottom types: silt, coarse sand and soft sedimentary rock; these materials span Chapman's range of observable ambient noise directionality characteristics.

6.6 Ship-generated Noise [6.4]

Whereas ambient noise has spectral characteristics which, over the operating bandwidth of a typical sonar, might well be described as uniform or "white", the same cannot necessarily be said of the acoustic emanations originating in machinery. In particular, our concern will be with the various propulsion machinery: propellers, motors, gearing and drive shafts associated with sea-going vessels. Other mechanical noise sources exist, however, and due attention may need to be addressed to them, in particular circumstances. Examples might include the high intensity shock waves encountered during piling operations for rig emplacement or other marine civil engineering purposes, or the variety of clanks, bangs, squeals and rattles associated with a wide range of loosely maritime activities, particularly in harbour and near-shore locations.

It must be stated that, whilst some general information concerning such noise sources is to be found [6.4, 6.5], an extensive (or at least publically accessible) database concerning source level, spectral content and spectral stability is not at this time available.

The propellers of a ship serve to generate noise in several ways. Firstly, propeller cavitation – a commonly encountered phenomenon manifest by the entrainment of air-bubbles into the wake – is the major cause of noise generated by surface ships. Propeller cavitation results from pressure fluctuations in the water in the vicinity (usually) of the blade tips. When these fluctuations fall below ambient pressure (effectively below atmospheric pressure, for a surface ship) dissolved air is caused to leave solution, forming the cavitation wake. Cavitation will exhibit a broadband, continuous (noiselike) spectrum peaking in the high tens of Hertz and falling at some 20 dB/decade, with increasing frequency. The spectrum may, however, be further confused by the presence of modulations at the propeller blade-rate. The magnitude of the spectrum level will in general increase markedly with increased speed. The propeller cavitation noise spectrum level of a surface ship, referred to 1 m standard range, may be estimated, in dB re 1 µPa per Hz, as a function of speed, v, in knots, frequency, f, in kHz and displacement tonnage T by using the empirical formula

$$SL = 10\log(3.10^3 \, v^6 T f^{-2}); \quad f > 1 \text{ kHz}$$

Secondly, a poorly designed propeller may exhibit "singing" when, typically, one blade-edge enters a high-frequency state of vibration whilst running. Such an effect may be most pronounced and will militate against quiet running. The problem may also be extremely difficult to cure, without redesign of the propeller itself. Finally, there may be expected a blade-rate tonal, with a frequency which is the product of the shaft speed and number of blades on the propeller itself. Virtually all modern merchant ships may be expected to generate shaft frequency tonals in the region 1–3 Hz and blade rate tonals with a fundamental in the range 6–10 Hz. The blade rate line series is the dominant feature of the low-frequency spectrum of ship-generated sounds. It does not follow that these tonals will increase in frequency with increasing speed, since thrust is usually controlled by using variable pitch propellers, with engine speed and thus shaft speed, maintained at an economical but sensibly constant level.

Noise deriving from reciprocating piston-engines, rotating (turbine) propulsion units and the various drive equipments associated with them is characterised by both a line and a continuous spectral content. In the medium and high-speed marine diesel engines which power many smaller vessels, piston-slap is the major source of noise. Piston-slap is the impact of the piston against the cylinder and is caused by the sideways movement induced by direction changes of the piston during the firing cycle. The reader may well be familiar with the observation that any diesel engine tends to run substantially more noisily when first started. This effect is primarily caused by excessive slap when the engine is cold; quietening occurs because the running tolerances tighten as the engine block warms up. Piston slap may occur several times in a single cylinder, during a single rotation of the crankshaft, and will be occurring also on all other cylinders in the engine. The result will be a rich harmonic line spectrum with a fundamental at the crankshaft rotation frequency.

Large vessels which incorporate large, low speed (<250 rpm) diesel engines will, however, not exhibit piston-slap as a significant noise-generation mechanism. This is a consequence of the design of articulated connecting rods, which virtually eliminate transverse motion of the piston.

Gears are also important sources of noise. In this case the tones generated will be at multiples of the tooth contact frequency and thus related to the product of drive shaft speed and the number of teeth on the driving gear. The drive shaft, by flexing or whipping during rotation, may also act as a source of mechanical noise with a line component, as may the shaft bearings themselves. Although ball-race bearings are in general noisier than friction (block) bearings, it is only when such bearings are poorly installed or approaching the end of their useful life that excessive noise will be generated.

Taking all these various effects into account, figure 6.9 illustrates the general form of the spectrum to be anticipated from a surface vessel, excluding tonals at the lower frequencies. It should be noted that tonals cannot be usefully incorporated on a graph such as this, since the vertical axis measures a spectral density in units of dB re 1 µPa per Hz of measurement analyser bandwidth. The tonals, being a spectral representation of a pressure sinusoid, are measured in dB re 1 µPa. Thus typical tonals for a large merchant vessel, the Chevron London, were identified at multiples of 6.8 Hz, with strongest components in the region 40 – 70 Hz, which corresponded to source levels (for each tone) of up to 190 dB re 1 µPa. The significance of this distinction lies in, for example, the ability of spectrum analysis equipments to discriminate tonals from cavitation noise. Taking the spectrum level in the 10 to 100 Hz decade, as indicated in figure 6.9 as being (about) 160 dB re 1µPa per Hz and guessing that we should need an analyser bandwidth of significantly less than the tonal spacing of 6.8 Hz – say 2 Hz – then the noise level due to cavitation noise will be

$$NL = NSL + 10\log B$$

where B is the effective analyser bandwidth, so that $NL = 160 + 10\log 2 = 163$ dB re 1 µPa, which compared against the tonal strength of 190 dB re 1 µPa indicates a signal-to-noise ratio of the order of 27 dB, which would clearly facilitate line identification. Notice that averaging times in building the spectral record would need to be significantly longer than the reciprocal of the analyser bandwidth – which is 0.5 second. Averaging over at least 5 seconds would appear to be necessary. In practice, it is probable that a digital (FFT) spectrum analysis on recorded time-series data would be used to accomplish such an analysis (see Chapter 3).

Because of the need to acquire and record frequency components down to 1 Hz, digital or FM tape recording of frequencies from (notionally) zero to perhaps 1 kHz would be taken on one channel of a multichannel recorder. Analysis could later proceed on this part of the spectral record using a 512 point real FFT, with suitably selected overlap, window weighting and

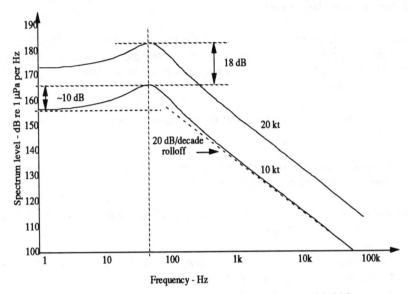

Figure 6.9. The noise spectral density for a large (30,000 tonne) merchant vessel

averaging, to yield the required 2 Hz of analyser bandwidth. Simultaneously, a full-width direct (analog) signal might be recorded on a second tape-recorder channel, holding frequency components to some high tens of kHz, but losing (because of the ac coupling of the direct recording format) the lower hundred Hz or so of the incoming signal. The second channel could then be analysed, but with coarser frequency resolution, a much higher sampling rate and the presumption that the effective analyser filter bandwidth was only some 100's of Hz, rather than the 2 Hz required to resolve the lower frequency spectral lines.

The various modulations on a surface vessel's acoustic output make its noise signature exceptionally distinctive, to the advantage of the experienced passive sonar operator intent upon target classification and identification. However, the reader should be aware that the measurement of spectral characteristics is at best difficult, if only because of the degrading influence of the acoustic channel separating a source vessel and a listener. The problem is, of course, most severe in shallow water. It should also be noted that, in such a context, "shallow" pertains to acoustic wavelength as much as geographical circumstance. At the tonal frequencies (1–30 Hz) corresponding to shaft and blade rate radiation, wavelengths are of the order of 100–1000 m. Under such circumstances, in the shallow shelf-seas, interference and waveguide effects will present experimental problems.

6.7 Reverberation [6.6]

All sonars, whether active or passive, are subject to the corrupting influence of noise. Active sonars give rise to another source of corruption known as reverberation, which is innately associated with several interlinking physical effects. These effects are:

1. Multipath propagation caused by boundary (sea-floor and sea-surface) effects.

2. Multipath propagation caused by a possible multiplicity of refractive ("mainpath multipath") transmission paths.

3. Volume scattering caused by suspended reflective and diffractive objects such as plankton and nekton.

4. Surface scattering, caused by sea-surface and sea-floor roughness or entrained air bubbles in the immediate surface layer.

The relative significance of these various causative phenomena will depend upon the circumstances under which an active sonar is required to operate. Their gross effect will be to cause a time-spreading of received signal energy. Furthermore, the reverberation intensity will increase, with increasing transmitter power. Both noise and reverberation can act to obscure a received signal. However, it is usually the case that one corruptive process will dominate. The sonar will tend to be either noise-limited or reverberation-limited. If the former is the case, then increasing the signal power will have the effect of improving the signal-to-noise ratio. If the sonar is reverberation

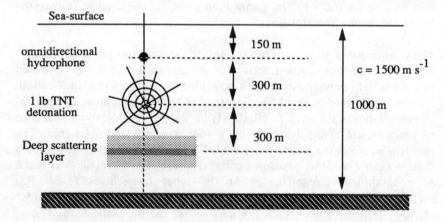

Figure 6.10. An experiment demonstrating the properties of surface and volume reverberation

limited, then no advantage will accrue from adopting such a strategy. In the main, the corruption induced by volume and surface scattering will, because of the large number of scattering entities, be largely incoherent. The same may well be true of "mainpath multipath". However, reverberation caused by multiple specular reflections from the sea-surface and sea-floor will cause attenuated and delayed replicas of the transmitted signal to be returned to the receiver.

To illustrate some of these various phenomena, consider the scenario depicted in figure 6.10. Here, we imagine a 1 lb charge of TNT to be detonated at a depth of 450 m, and thus 300 m vertically below a hydrophone at a depth of 150 m. The water depth is presumed to be (about) 1000 m.

Some 200 ms after detonation, a first pressure shock-wave will arrive at the hydrophone. This wave will have the temporal characteristics of an explosive detonation, exhibiting a rapid rise to a peak pressure of the order of 230 dB re 1 µPa, followed by a slower, noisy, envelope decay of intensity caused by bubble formation and collapse. The time-constant of this decay will be about 200 µs for an explosive charge of the size envisaged here. The next event experienced by the hydrophone is the passage of the surface reflection and the associated surface reverberation. The specular surface reflection will dominate, in magnitude, at the outset. The reverberation will persist longer, however, than the nominally three time-constants taken for the specular component to decay to a negligibly small level. Notice that, as the reflection passes the hydrophone, the upper rim of the reflected spherical region of high pressure will be travelling circularly outwards within the surface layer of the sea, inducing a reverberant return which will be substantially omni-directional, if not exactly isotropic – or of equal intensity in all directions – as figure 6.11 suggests. We shall thus expect the hydrophone to respond first to the reflected pressure-wave, then to the reverberant surface scattering return.

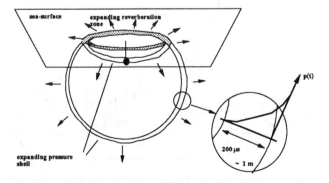

Figure 6.11. The expanding high-pressure shell following a submarine detonation

Some 200 ms after the detonation, the expanding pressure shell will intersect with the deep scattering layer. This layer contains a profusion of planktonic animals and, feeding upon them, a range of nekton often particularly characterised by species of squid. In any event, a strong volume reverberation is to be anticipated, arriving at the hydrophone (as a backscatter signal) about 600 ms after the detonation.

The shock-wave will progress outwards until it reflects from the ocean-floor. A specular reflection will follow, which will initiate a response from the hydrophone approximately one second after detonation. Associated with this reflection, but following it, will be a further bottom reverberation. Yet further multiple surface and bottom reflections, all with associated volume and surface reverberation, may be expected, decaying – of course – as thunder from a lightning-strike rolls away between the hills surrounding a valley. All these various events are depicted in figure 6.12, which provides an indication of both the time-scale and magnitude of the various events contributing to produce a reverberant response to a pulse of high energy but short duration.

6.8 Scattering [6.7]

Scattering, as we have seen, takes place in such a manner that sound, incident upon a scattering surface or within a scattering volume, is re-radiated in directions other than that which would correspond to specular reflection. Because active sonars are often monostatic (the transmit and

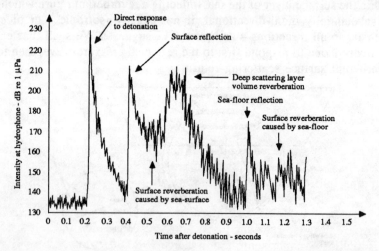

Figure 6.12. Response at the hydrophone to an explosive detonation. The explosion time constant (here about 200 ms) and peak amplitude (230 dB re 1 µPa) both increase with increasing charge size

Noise and Reverberation

receive transducers are at the same location) much attention has been given to the problem of backscattering, where the intensity of sound returning on the same path as the transmission is the quantity which is required to be estimated. However, there has been growing interest in bistatic geometries, in particular for communication purposes where, necessarily, the transmitter and receiver will be at different geographical locations, and also in passive sonar applications. This latter class of problem stems from a military objective: the covert detection and recognition of surface or submarine targets. In both these examples, the receiver is required to isolate a main-path received signal from reverberation and noise. The reverberation may now contain a significant forward-scattered component, and backscattering is likely to be quite unimportant. Since most available data pertain to backscatter, some care will be needed in acquiring modelling information for system performance prediction.

We turn our attention first to the problems of volume and surface backscatter prediction. We note that the performance criteria we shall investigate pertain specifically to pulsed sonars and would not be accurate for continuous wave transmissions, which have received relatively little study in the scattering context. Because we deal with pulsed sonar, let us examine, for a moment, the temporal and spatial characteristics of the pulse. A sonar pulse of duration T seconds will clearly correspond to a pressure perturbation in the water of cT metres spatial extent. Imagine such a pulse to enter a region of scatterers, such as the deep scattering layer. Imagine also the layer boundary to be reasonably precisely defined. The pulse front edge will become incident upon the layer boundary and will begin to produce backscatter which will proceed backwards towards the transmitter. The front edge will penetrate the scattering volume, initiating backscatter as it travels inwards. Some time later (depending upon how far the layer is from the transmit-receive transducer location) backscatter will begin to be detected by the receive transducer. The actual acoustic contribution at any instant will be the sum of backscatter initiated by all parts of the pressure perturbation which is the travelling pulse, which are at that instant equally time-spaced from the receiver.

Put in another way, the front end of the pulse will have stimulated a scatterer to return an energy contribution towards the receiver. When the pulse is "half-way into" the layer, it will stimulate a backwards travelling pulse *which will have got back to the layer boundary at exactly the same time that the pulse rear edge will also have got to the boundary, travelling inwards.* The pulse rear edge will thus, at that instant, be initiating an incoherent backscatter contribution which will be power additive with the backward travelling contribution from the front edge of the pulse. Of course, the same effect will be occurring for all other parts of the pulse. It is as if, *at that instant*, a layer depth of one half the spatial length of the pulse were contributing, power additively, to form the backscatter wavefront about to proceed back to the receive transducer.

Having thus established the spatial extent of the scattering volume as $cT/2$, we may now consider the geometry of a typical volume backscatter situation, figure 6.13. For simplicity, let us assume a transmit–receive transducer with an axially symmetric beam-pattern, and a beam-width of ϕ radians. We discuss the evaluation of ϕ in terms of transmit frequency and transducer aperture in Chapter 8. At range R, the subtended circular radius of the scattering volume will be ϕR, and the scattering volume itself will be $\pi\phi^2 R^2 cT/2$. If the transmitted intensity is I_o, then the intensity incident on the scattering volume will be $I_o R^{-2}$. We assume a backscatter constant κ_v, determining the proportion of incident energy per unit scattering volume, returned in the direction of the transmitter, and calculate, subject again to a further inverse square-law spreading, a received intensity given by

$$I_1 = \kappa_v \, I_o \, \pi\phi^2 R^{-2} cT/2$$

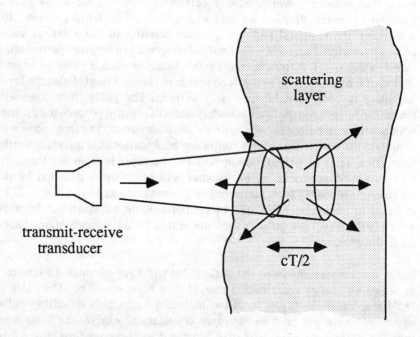

Figure 6.13. Volume backscattering: the geometry of the problem

This relation may be used to predict the backscatter strength, if κ_v is known for a given ocean environment. Alternatively, it may be re-cast to allow κ_v to be estimated from measurements of I_0 and I_1.

We further note that the backscatter intensity increases with the source intensity, decreases as the square of range to the scatterer location, increases as beamwidth increases and also increases with increasing pulse length. As was mentioned in an earlier paragraph, the formula derived above is good for "short-pulse" sonars. Such sonars would encompass some tens or hundreds of cycles of carrier within the pulse envelope.

A good example of volume backscatter is to be found in the deep scattering layer, referred to above in the context of observations of reverberation produced by a small explosive detonation. The depth and population dispersion of the deep scattering layer is subject to wide diurnal variation. It might typically be found at depths of less than 200 m during the night, making a rapid movement to and from depths in excess of 1000 m at sunrise and sunset. The population concentration at the surface at night is such as to produce a value of $10\log \kappa_v$ of about -70 dB. During the day, the value of $10\log \kappa_v$ falls to between -90 and -100 dB. Finally, the backscatter coefficient shows little observed frequency variability at frequencies in excess of 20 kHz. The use of backscatter as a method of determining biomass concentration is now a matter of considerable scientific interest, and is a subject to which we return, in Chapter 9.

Another example of volume backscatter is to be encountered when a sonar is made to impinge on the wake of a vessel. Because the wake is heavily entrained with air-bubbles which persist long after the passage of the vessel itself, a strong backscattered return may be observed for many minutes. The wake strength, measured in decibels per metre of wake length insonified, replaces the scattering strength coefficient $10\log \kappa_v$. Typical values of the order of -10 to -30 dB at frequencies of some tens of kHz have been observed, higher values corresponding to larger vessels. The wake strength decay rate is of the order of 1 dB per minute. Much information on the subject of wakes and their acoustic properties is to be found in reference [6.8].

A similar argument to that presented above for volume scattering may be applied to show that, for surface scattering, the backscatter expression is

$$I_1 = \kappa_s I_0 \pi\psi R^{-3} cT/2$$

where ψ is the beamwidth of the transmit–receive transducer, in the horizontal plane, and κ_s is a surface reverberation constant. The major causes of sea-surface backscatter are surface roughness induced by wave action and backscatter produced by air bubbles in the immediate surface layer.

Urick [6.9] discusses various theoretical and empirical formulae which have been suggested to allow calculation of the surface scattering coefficient κ_s. However, the unification of the available data to provide anything approaching a general formula $\kappa_s(\theta,f,v)$, which is of widespread applicability, yet remains wanting. Here, θ is incidence angle, v is windspeed and f the transmission frequency. Indeed, the more general problem of quantifying the scattering coefficient for arbitrary incident and observation angles, θ and ψ respectively: $\kappa_s(\theta,\psi,f,v)$, is even more elusive.

Surface scattering at the sea-floor could likewise well afford some further study and unification of results. The general problem has been likened to that of the diffuse reflection of light from a matt surface, which is described by Lambert's Law. The acoustic analog of Lambert's Law may be stated thus

$$I_1 = \mu I_0 \sin\theta \sin\psi$$

where I_1 is the intensity at unit distance from a unit area of diffuse scattering surface of reflection coefficient μ, insonified by plane waves of intensity I_0. Note that θ and ψ, the incidence and observation angles respectively, need not be co-planar. The backscatter coefficient (for which $\psi = \theta$) is thus

$$\kappa_s = \mu \sin^2\theta$$

Lambert's Law fits, by observation, reasonably well to some types of sea-floor, particularly those that are rough by comparison with the wavelength of the sound incident upon them. The value of μ may be bracketed between $\mu = 0.32$ and $\mu = 0.002$. The former value is the maximum reflectivity if the scattering is omnidirectional and no sound loss into the sea-floor takes place. The smaller value corresponds to experimental observations in the deep ocean [6.10, 6.11].

References

[6.1] G.M. Wenz, Acoustic Ambient Noise in the Ocean: Spectra and Sources, *J. Acoust. Soc. Am.*, Vol. 34, No. 12, 1962, 1936-1956

[6.2] R.J. Urick, *Ambient Noise in the Sea*, Peninsula Publishing, Los Altos, Calif., 1984, pp. 3-5

[6.3] D.M.F. Chapman, Surface Generated Noise in Shallow Water: A Model, *Proc. Inst. Acoustics (London)*, Vol. 9, Pt 4, December 1987, pp.1-11

[6.4] D. Ross, *Mechanics of Underwater Sound*, Pergamon Press, New York, 1976

[6.5] R.J. Urick, *Principles of Underwater Sound for Engineers*, McGraw Hill, New York, 2nd edition, 1975, pp.298-342

[6.6] *Physics of Sound in the Sea - Part II: Reverberation*, US National Research Council (Originally issued as Division 6, Volume 8 NRDC Summary Technical Reports)

[6.7] R.P. Chapman, Sound Scattering in the Ocean, in *Underwater Acoustics*, Vol. 2. (V.M. Albers, ed.), Plenum Press, New York, 1967, pp. 161-183

[6.8] *Physics of Sound in the Sea - Part IV: Acoustic Properties of Waves*, US National Research Council (Originally issued as Division 6, Volume 8 NRDC Summary Technical Reports)

[6.9] R.J. Urick, *Principles of Underwater Sound for Engineers*, McGraw Hill, New York, 2nd edition, 1975, pp. 240-243

[6.10] K.V. Mackenzie, Bottom Reverberation for 530 and 1030-cps Sound in Deep Water, *J. Acoust. Soc. Am.*, Vol. 33, No. 11, 1961, pp.1498-1504

[6.11] P.B. Schmidt, Monostatic and Bistatic Backscattering Measurements from the Deep Ocean Bottom, *J. Acoust. Soc. Am.*, Vol. 50, No. 1, Pt 2, 1971, pp. 326-331

7 Acoustic Transduction

7.1 Introduction

Underwater sound transducers convert electrical energy into or from mechanical energy, the latter quantity being perceived as longitudinal pressure waves in water. We thus impress a voltage waveform v(t) on the terminals of a transmit transducer, or projector, and generate a sympathetic pressure fluctuation p(t) in the water. The reverse occurs with a receive transducer, or hydrophone.

For acoustic projectors we seek, in partial characterisation, a projector constant k_p measured in terms of the logarithmic ratio of generated intensity per volt rms applied to the projector terminals. That is, k_p will have "units" of [dB re 1 µPa v^{-1}]. More specifically, the intensity so used to characterise the projector will be the on-axis intensity, reduced to a standard range of 1 m.

In making these further stipulations, there is a presumption that any projector will have some polar response (about which we shall concern ourselves shortly) and that calibration measurements (by whatever means) will be executed "in the far field" at some suitably large range R. Yet further presumptions would require that, not only was the calibration measurement far-field, but that it was also "free-field", so that reflections of sound to, for example, a calibrated test hydrophone from the sea-surface or sea-floor, or the walls of an acoustic test-tank, would not render the calibration inaccurate. If our calibration measurement thus yielded an on-axis intensity I at range R, then we should expect an intensity IR^2 at the reduced range of 1 m. In estimating the extent of the near-field and thus the range at which we might expect far-field conditions to start, we use the empirical result $R_c = A/\lambda$ where A is the projector surface area and λ is wavelength in water (recall that $c = f\lambda$ where f is the transmission frequency in Hz).

For the hydrophone, by contrast, the calibration constant k_h represents rms terminal voltage generated by immersing the hydrophone in a pressure field of given rms pressure. k_h thus has units of [V µPa^{-1}]. The same comments regarding the practicalities of calibration remain, of course. Commonly, a reference hydrophone (a "secondary standard" carefully calibrated by a manufacturer with access to good calibration facilities) will be used as the basis of a "comparison method" of further calibrating both acoustic projectors and equipment hydrophones. For the former, rms terminal voltage at range R from the projector will be used to infer rms pressure in the water at the hydrophone location caused by driving the projector with a known rms terminal voltage. This pressure will then be used to calculate acoustic intensity at the hydrophone. Finally the intensity at range R will be reduced to the intensity at 1 m standard range, and the calibration will be complete.

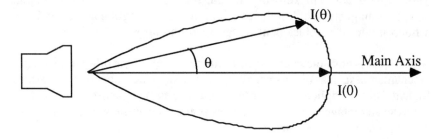

Figure 7.1. Projector polar response

For the hydrophone calibration, straightforward substitution of equipment hydrophone for reference hydrophone in a pressure field remote from any suitable projector will allow relative sensitivity at a given transmission frequency to be evaluated. A wider range of calibration methods is discussed in some detail in Urick [7.1].

In discussing projector calibration, the projector polar response was referred to. We write $b(\theta)$ as the ratio of off-axis to main-axis intensity, figure 7.1, so that $b(\theta) = I(\theta)/I(0)$. The "3 dB" half-beamwidth (θ measured away from the main axis, until $I(\theta)/I(0) = 0.5$ or $10 \log_{10}(I(\theta)/I(0)) = -3$ dB) will often be of particular interest in estimating the suitability of a projector for a particular task. For a circular transducer of diameter D, the half-beamwidth measured in degrees will be approximately $30\lambda/D$ where, again, λ is the wavelength of sound in water. Similar polar response considerations apply to the hydrophone.

7.2 The Basic Principles of Acoustic Transduction

The objective of any transduction operation is to bring about the conversion of energy from one physical form to another. Often, because of the wide use of electronic methods for data gathering and message transmission, the transduction will be to or from the electrical state. Thus for underwater acoustics applications we often seek physical phenomena which will translate between the electrical state and some form of mechanical displacement, since the latter will then engender pressure fluctuations in the water.

The most common form of transduction, and the one upon which we shall primarily focus our attention, makes use of the piezo-electric effect. However, it should be noted that magnetostriction is of great importance in the design of some forms of transducer, particularly for low-frequency applications. An example of a scroll-type flooded ring transducer is shown in figure 7.2. The scroll is made of nickel-iron alloy and the application, via the surrounding coil, of a magnetic field, induces circumferential length changes and thus a hoop mode of displacement of the surrounding water.

Although it is possible to envisage applications involving other transduction mechanisms, such as piezo-resistivity or displacement capacitance (perhaps for hydrophone applications), the piezo-electric and magnetostrictive methods have the advantage of extreme robustness and relative immunity to pressure in deep water applications.

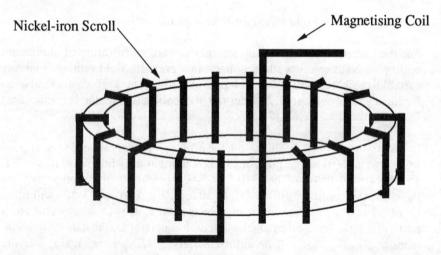

Figure 7.2. Free-flooding, low-frequency magnetostrictive transducer; diameter ~ 1 m

It is often interesting and sometimes helpful, in the context of instrumentation transduction, to envisage the transduction process as being composed of several separable stages. Thus for example, although the strain gauge diaphragm might be thought of just as "a transducer", it should more correctly be seen as a diaphragm transducer, converting pressure fluctuations to mechanical displacement, followed by the strain gauge, converting displacement to resistance by a body-distortional mechanism. Such a visualisation allows greater insight into the role played by the various components involved in the transduction, and how they may individually affect the sensitivity, linearity or inherent accuracy of the process itself.

Usually, acoustic transduction is a rather more "rough and ready" job than instrumentation transduction. However, there is at least one relatively new transducer which closely parallels the diaphragm displacement pressure sensor described above, albeit in reverse – to generate pressure fluctuations rather than detect them. That device is the flextensional transducer, figure 7.3, wherein an elliptical containment provides a diaphragm-like displacement across the minor axis, when the piezo-electric drive stack produces an extension or contraction along the major axis. The elliptical form also confers a leverage action (a circular container would not), increasing the face displacement and thus the amplitude of the pressure wave launched into the water, thereby assisting matching and providing a relatively broadband transduction operation.

Finally, it is interesting to note that, whereas piezo-electric transduction and, to a lesser extent magnetostrictive transduction remain the workhorses in so far as transducer design is concerned, there yet remains a wide range of mechanisms primarily, but not exclusively, utilised for marine seismic survey, which include explosive and air-gun detonations and electric spark discharges.

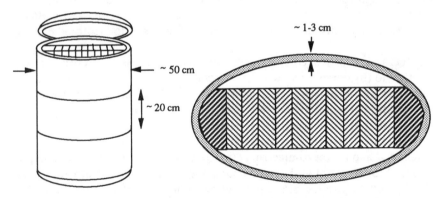

Figure 7.3. The flextensional transducer, showing use in a short stave of three flextensional units. Also shown are the drive stack of piezo-electric elements, metal endpieces and an aluminium or fibreglass shell. Operating frequency is in the range 500 Hz to 3 kHz depending upon wall thickness and material

These mechanisms are used to launch high-intensity, low-frequency shock waves into the water and, particularly, into the underlying geological strata.

7.3 Piezo-electric Transduction

The phenomenon of piezo-electricity lies at the heart of the acoustic transduction process. Of the thirty-two classes of crystal structure, about two-thirds exhibit the property that charge separation, and thus the generation of a transverse electric field, occurs with mechanical displacement along appropriate crystal axes. Quartz is well-known to exhibit the effect, and its various physical properties make it most suitable as a stable, frequency-defining oscillator element. In the past, quartz and, indeed, a range of other materials were employed in the construction of underwater transducers. At the present time, most transducers make use of piezo-electric ceramic materials and of these, the lead zirconate titanate (PZT) ceramics are by far the most commonly encountered. By analogy with the phenomenon of ferromagnetism, the piezo-electric ceramics are often referred to as "ferroelectric" materials, despite the fact that iron is not a constituent of their molecular structure.

Although, of the possible piezoelectric transducer materials, ferroelectric ceramics are now used exclusively for projector transduction, a small proportion of receiver applications do make use of a plastic film known as poly-vinylidene fluoride (PVF). PVF is sensitive as a receiver, but exhibits high capacitance and is a poor projector material for practical as well as physical reasons. Unlike the ceramic materials, it is well-matched to water and is thus an efficient converter of acoustic to electrical energy.

PZT ceramic is produced as a powder which is compressed and fired to form brittle and (by comparison with other ceramics) soft solids, which typically take the form of rings, discs, tubes, plates and half-spheres. Further working by cutting, using a diamond saw, and by grinding and lapping to produce special shapes with preferred acoustical properties, is also possible. As manufactured, the fired ceramic is un-poled: its crystal structure is un-orientated and no piezo-electric action is discernible. By way of example, a disc of the material will have its opposing flat faces silvered, with a heat-fired paint, to provide plane-parallel circular conductive electrodes. The crystal will be placed in a high electric field and heated to a temperature in excess of some 300° C. This is the "Curie temperature" above which piezo-electricity is destroyed, the crystal structure having been "loosened up" because of the heating. The high electric field has the effect of pulling the electric dipoles within the crystal, to make them lie with their dielectric axes along the lines of electric flux. As the crystal is cooled down, still with the high electric field applied, the crystal dipole orientation becomes "frozen" and, reduced to room temperature, strong piezo-electricity will be evident in the disc. Pressure applied across the silvered flat faces of the disc will result in the generation of possibly sizeable electric fields and, even, kilovolts of terminal voltage. By contrast, connecting an ac signal generator to the disc faces will, at audio frequencies and modest voltages, produce an audible whistle.

Ferroelectric ceramics are materials which have a high acoustic impedance, of the order of 35 MRayl, and low internal mechanical loss. The transmission coefficient between the material and water is given by $T = \rho_2 c_2/(\rho_1 c_1 + \rho_2 c_2)$ which thus has a value of about 0.04. Consequently, if an element is caused to vibrate, by means of an applied alternating voltage, and thus establish an internal acoustic wave motion, very little of the sound emanates into the water. This has two consequences. Firstly, in order to get useful amounts of acoustic energy into the water, high drive voltages are required. Secondly, if excited with an impulse, the transducer (like a struck cymbal) "rings" sustainedly. The time domain "envelope" is thus of long duration , and consequently the transducer pass-band is small. The transducer element, if impulsively excited, will ring at a frequency determined by its width, h, and by the speed of sound within it.

In order to improve the transfer of sound into the water, it is possible to provide a matching layer of acoustic impedance which is intermediate between that of the ceramic and that of the water. The optimum layer impedance is about 13 MRayl. No engineering materials quite provide such an impedance. However, aluminium, titanium and loaded epoxy resins come usefully close. In the following sections we consider in greater detail the structure and design of a variety of composite transducers which use ferroelectric ceramics as the driving elements.

7.4 The Langevin Projector

Any of the naturally occurring piezo-electric materials and even, from the viewpoint of economy, the PZT ceramics, are not readily to be had in objects of sufficiently large physical dimension to produce resonant behaviour in the audible range. Many sonars are required to operate at such frequencies and this was particularly the case in the early days of practical electronic sonar, after World War I. In 1920 Paul Langevin published [7.2] a patent disclosure where he described how mechanical structures sandwiching quartz driving elements might be employed for the resonant generation of sound. The simplest Langevin resonator is shown in figure 7.4.

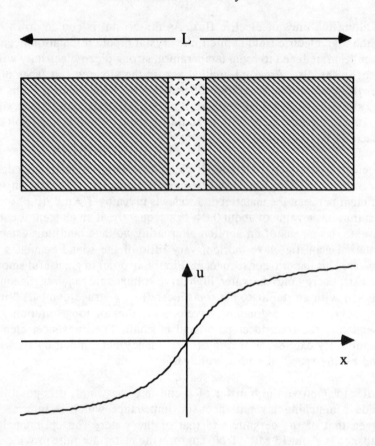

Figure 7.4. The Langevin resonator

Here, a PZT disc is imagined to be glued to cylindrical end-bars of, perhaps, a metallic material with a similar acoustic impedance and sound speed as that of the PZT itself. For example, the various PZT-type ceramics exhibit acoustic impedance in the range 25–35 MRayl. Brass has an acoustic impedance of about 30 MRayl. Metals such as brass, steel and the various aluminium magnesium and titanium alloys have high quality factors (even by comparison with PZT) and thus make potentially good resonator materials with low internal conversion of acoustic energy to heat. The entire mechanical structure is "driven" by the central crystal and will resonate at a frequency f' dictated predominantly by the length, L, of the structure according to a law $L = \lambda'/2 = c'/2f$ where λ' is the wavelength of sound in a material of sound speed c'. The graph in figure 7.4 shows "particle" velocity, u, versus axial displacement, x. The end-faces make the greatest velocity and spatial excursions.

Even designing a simple resonating structure such as is shown in figure 7.4 involves consideration of some less obvious properties of sound propagation in solids. The speed of sound in a thin wire, often referred to as "bar velocity", v_{bar}, is higher than the velocity in bulk materials, v_{bulk}. The effect is a consequence of the bulk displacement nature of vibrations, wherein a lateral contraction gives rise to a thickness increase in, for example, a stubby bar. This effect is described by Poisson's ratio, which for most metallic and piezo-electric ceramics of interest is about 0.3. Figure 7.5 illustrates the effect [7.3, 7.4], which is most marked for "squat" transducers.

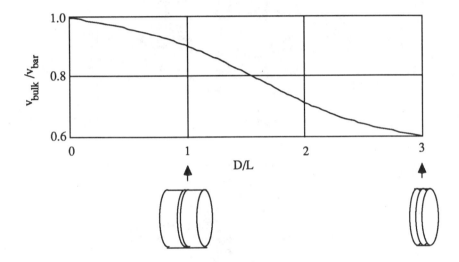

Figure 7.5. *Normalised phase velocity for extensional waves in round bars; Poisson's ratio: 0.3*

We may now take the design of the Langevin projector through a series of stages which will allow us to develop it into a rather more practical underwater projector transducer. The first three stages are of a practical nature. If we use two or, indeed, any even number of driver PZT discs, then interconnection and orientation may be made so as to add end displacement and parallel the discs in terms of their electrical loading. The method is illustrated in figure 7.6. Metal shims between the discs provide electrical connection. The two metallic endpieces are commoned and "earthy". This is particularly convenient when the discs are replaced by PZT rings and a central "pre-stress" bolt is used to strengthen the whole structure, figure 7.6. The shims may be beryllium copper gauze or thin sheet or even 0.5 mm sheet stainless steel, without unduly reducing efficiency. Yet another feature of the design may be appreciated by noting, in figure 7.4, that the

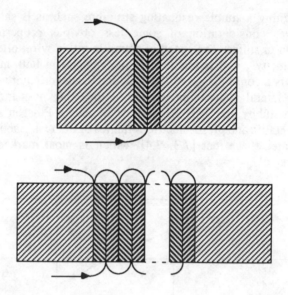

Figure 7.6. Development of the Langevin resonator (1)

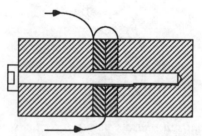

Figure 7.7. Development of the Langevin resonator (2)

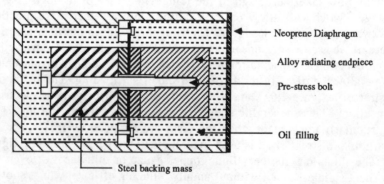

Figure 7.8. Development of the Langevin resonator (3)

vertical centre-line is a nodal point and thus, potentially at least, a good place at which to attempt a mechanical mounting of the device, since the mount will not then dampen vibration. This leads us through to the next phase of design, where we note first that the coupling between PZT and water via brass ends (assuming our first resonator to be so formed) is poor. The coupling is best achieved between materials such as PZT and water, with an endpiece material the specific acoustic impedance of which is the geometric mean of the impedances of the materials which it separates. For PZT and water (with acoustic impedances of 1.5 and 30 MRayl respectively), we anticipate an endpiece material impedance of about 7 MRayl. Magnesium alloy has an acoustic impedance of about 8 MRayl and so would do well, at least in this one respect. Aluminium alloy exhibits an acoustic impedance of about 14 MRayl.

Although this would produce a severer mismatch, such a transducer would still work well. The design method is forgiving. If the tailpiece is actually made of a material with a higher acoustic impedance than PZT, then coupling into a fluid at that end will be yet poorer than from PZT alone. We may thus evolve a design such as is illustrated in figure 7.8, where the entire resonant structure is housed in an oil bath, the oil having much the same acoustic impedance as water, and separated from the water by a tough neoprene rubber membrane.

In fact, the design may be radically simplified for some applications, albeit at the possible cost of transducer efficiency, by using a polyurethane rubber

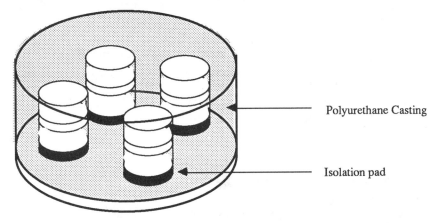

Figure 7.9 Use of the Langevin resonator in multiple transducer arrays

encapsulation technique, figure 7.9. If an array of resonators is encapsulated in this way, in an attempt to increase the effective aperture and reduce beamwidth, then resonator/resonator and resonator/backplate interaction may also prove troublesome.

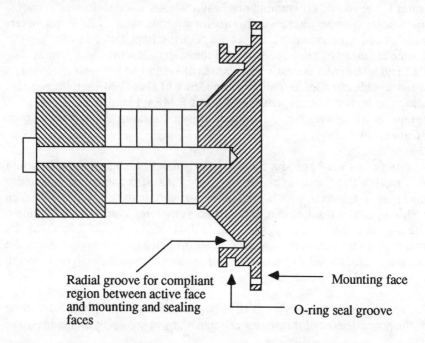

Radial groove for compliant region between active face and mounting and sealing faces

Mounting face

O-ring seal groove

Figure 7.10. The Tonpilz resonator

The final stage of development of the Langevin resonator is into what is often termed the "Tonpiltz" resonator, figure 7.10. Tonpiltz means "sound mushroom" and would appear to be World War II terminology adopted from German Naval parlance. In any event, the term is descriptive of the structure itself, wherein the radiating face is flared outwards, again to improve the matching into the water load.

7.5 Ring and Tube Transducer Designs [7.5]

Our objective thus far has been to develop transducer designs which allow relatively little ceramic material to act as the driving element for a larger mechanical body with a relatively low resonant frequency. Another method of attaining the same end with possibly less technical complication is to utilise a free-flooding ring transducer. This, figure 7.11, simply consists of a short, large-diameter, thin-walled ceramic tube operating in a hoop resonance mode.

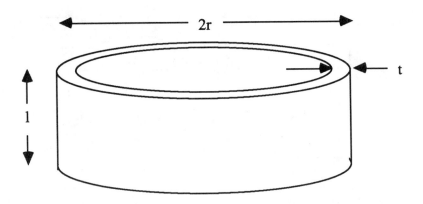

Figure 7.11. The free-flooding ring

Similar structures have been used fully encapsulated in polyurethane, rather than dip-coated to be free-flooding. Smaller, and thus higher frequency, tubes have also been discussed in the literature [7.6] as transmitter devices of simplicity and robustness. It is, perhaps, appropriate at this stage to summarise the hoop, radial, length and thickness resonances for the tube transducer, since this ceramic shape is readily available and extremely useful for a wide range of transmission and reception applications.

The radial mode resonance is given as $f \approx c/2\pi r$, the length mode resonance by $f \approx c/2l$ and the thickness mode by $f \approx c'/2t$ where c is the velocity of longitudinal waves, c' the velocity of thickness waves, r the mean radius, l the length and t the wall thickness of the tube.

7.6 Resonance Behaviour of Transducers

If the development of the various transducers described above has been correctly carried out, some care will have been taken to ensure that the transmission resonance is kept reasonably separated from other dominant resonances. Then the transducer will act as a single resonator and will have a terminal electrical response which will equate to that of a series tuned LCR circuit. Because the drive elements are ceramic rings with fired plate electrodes, and because ceramic is a dielectric material, there will be, placed in parallel with the LCR equivalent of the mechanical resonator, a capacitor of significant value, which models the static capacitance, C_0, of the drive elements themselves. The equivalent circuit will then have the form shown in figure 7.12. Here the resistive component in the series LCR circuit has been split into two components: the radiation resistance R_r and the loss resistance R_l.

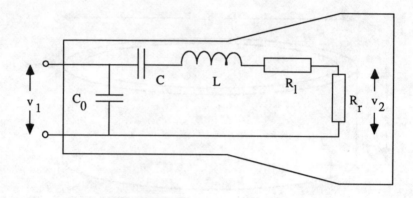

Figure 7.12. Equivalent circuit of the single resonance transducer

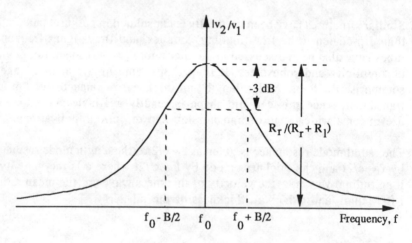

Figure 7.13. Transmitting response of the untuned single resonance transducer

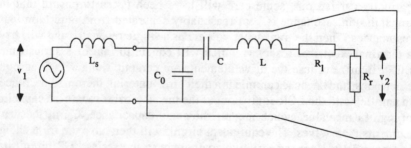

Figure 7.14. The single resonance transducer with series tuning

We are interested first in finding the overall voltage transfer function, v_1 which, since v_2 is proportional to the pressure field developed in the water, gives the effective shape of the transmitter response, figure 7.13. It is easily shown that

$$|v_2/v_1| = \omega_0 C R_r \{(\omega/\omega_0 - \omega_0/\omega)^2 + Q_0^{-2}\}^{-1/2}$$

$$= R_r/(R_r + R_l) \text{ if } \omega = \omega_0$$

where $\omega_0 = (LC)^{-1/2}$, $Q_0 = (\omega_0 C(R_r + R_l))^{-1}$ and the bandwidth, measured in Hertz, is given by $B \approx \omega_0/2\pi Q_0$ if $Q_0 \gg 1$. This function is not a particularly attractive transmitter response from, at least, a sub-sea communications standpoint. By employing series tuning, figure 7.14, a double-humped response can be obtained, figure 7.15, although tailoring the depth of ripple may be tricky, in practical terms.

Note that the one remaining variable which is not, as it were, totally buried within the transducer structure and thus difficult to gain access to without a complete structural re-design, is the static capacitance, which may be padded-up to a larger value. This will inevitably have the effect of drawing together the two humps in the transmission characteristic, increasing the overall "gain" and reducing the ripple, but at the expense of bandwidth. In many sub-sea communications applications, where tailoring the transmission response in this way might be desirable if data rate is not at a premium, this might not be a problem.

There remains one further possible objective in performing series tuning. Given a Langevin or Tonpilz-type transducer, the Q-factor may only be modest and the damping on series tuning could significantly undermine any Q-magnification or transformer action in "stepping up" the drive voltage levels issued by semiconductor drive circuitry. Remember, also, that acoustic transducers often present terminal resistance, when properly tuned to cancel the quadrature current drain demanded by the static capacitance, of some hundreds or even thousands of ohms. To drive powers of the order of tens or hundreds of watts into such transducers may require drive voltages of hundreds of volts or even kilovolts. For the Langevin and Tonpilz transducers, this means that a transformer must be interposed between the source of the drive waveform and the terminal input to the transducer, including the series tuning inductor. In contrast, for the flooded ring or for any piezo-electric crystal left deliberately unmatched mechanically, where the Q-factor is intrinsically rather high, series tuning can lead to a worthwhile transformer action, albeit with relatively small available transmitter bandwidth. In some applications this may not matter.

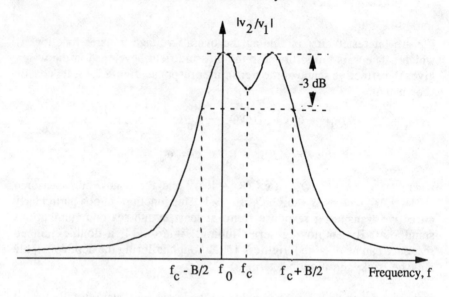

Figure 7.15. Transmitting response of the tuned, single resonance transducer

At series resonance, the inductive and capacitive reactances due to L and C in figure 7.14 cancel, so that the total resistance $R_r + R_l$ appears across C_0. The overall voltage gain when $\omega = \omega_0$ then becomes

$$|v_2/v_1| = \{R_r/(R_r + R_l)\}\{(R_L/(R_r + R_l))^2 + L_sC(R_L/L_s) + (1/(C(R_r + R_l)))^2\}^{-1/2}$$

where R_L is the self-resistance of the tuning inductance, L_s. If $R_L \to 0$ then $|v_2/v_1| \to \omega_0 C R_r$. If $R \to \infty$ then $|v_2/v_1| \to (\omega_0 L_s/R_L)(R_r/(R_r + R_l))$. Both extremes represent Q-factor gains which might thus be expected to offer an effective transforming action with a voltage step-up of about ten. It is also worth noting that, in this mode of tuning, the two resonance peaks which inevitably occur when a series inductor is added are widely separated, not close together forming a notionally flat, passband characteristic, such as was shown in figure 7.15.

Finally, it is interesting to reflect on the design requirements imposed upon the series tuning inductor and on any parallel, padding capacitance, required to increase the value of the static capacitance, C_0. First, the tuning inductor must not saturate, or even approach saturation. This can be quite a difficult objective to meet, and it is as well to be prepared for difficulties, particularly when tuning modest to low frequency transducers. If saturation is even approached, the non-linear behaviour of ferrite materials leads to significant "describing function" phase shifts and consequent departure from "inductor-

Acoustic Transduction

like" behaviour. The result will be a dramatic and often mysterious detuning and loss of performance. Curiously, choosing adequate capacitors can be extremely difficult, also. Here the problem is that the terminal voltage across the transducer, and hence across any additional padding capacitance, may well be of the order of kilovolts. Typically, capacitors are specified according to a dc voltage rating, which may well be only one half or one third the effective safe minimum ac rating. Since it is quite difficult to find suppliers of capacitors rated for even dc operation to 1 kilovolt, never mind find a complete range of preferred values, the problem will be readily appreciated.

An interesting further consequence to the series tuning problem has been investigated by Coates and Mathams [7.7]. A moment's thought will show that the transducer equivalent circuit of figure 7.13 is capable of being thought of as the terminating end of a T-filter section. If, as figure 7.16 suggests, we increase the complexity of the matching network, we may hope to fabricate transducer matching networks capable of realising some, if not all, of the all-pole modern passive filter designs. It is even possible to treat the mechanical end of the transducer equivalent circuit in a similar way, as Dunn and Smith [7.8] have recently pointed out. Considerable potential thus exists for tailoring a transmission response to desired passband criteria.

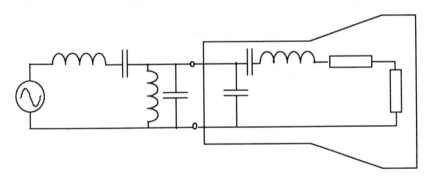

Figure 7.16. The matching circuit approach

7.7 Multiple Matching Layer Transducers

Transducers of the Langevin and Tonpiltz type are predominantly the most popular designs for operation in the broad range of frequencies from (about) 1 kHz to (about) 100 kHz. For frequencies in excess of 100 kHz, up to (for reasons related only to the practicalities of fabrication of robust, well-engineered transducers) perhaps a few MHz, another approach to design becomes feasible. At such frequencies, it becomes relatively easy to obtain drive elements (typically round PZT discs) which are of large diameter by comparison with their thickness: the so-called "thin disk". Such drive elements, figure 7.17, if placed directly in contact with a water load, and assuming an air backing, will exhibit a high Q-factor (typically about 15) and a Lorentzian (single tuned circuit) transmission response into water, of the form already illustrated in figure 7.13. The transmission band centre frequency (the resonance frequency of the disc) will be $f_0 = c'/2d$ where c' is sound speed in the disc (typically about 3000 m s^{-1}) and d is the disc thickness.

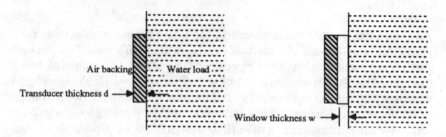

Figure 7.17. *The thin disc transducer and its matching into water*

If an acoustic quarter-wave matching layer, of thickness $w = dc''/2c'$ where c'' is sound speed in the matching layer, and acoustic impedance equal to the geometric mean of the acoustic impedances of PZT and water respectively, is interposed between the PZT disc and the water load, improved matching will result. The bandwidth will broaden, and the transmission response will exhibit, as in figure 7.15, a double humped bandpass characteristic. Some difficulty attaches to finding a suitable matching layer material at the required impedance of about 7 MRayl. Curiously, the author – following a hint in Wells [7.9] has successfully made up composite samples of mammalian long-bone which could be turned into matching layer discs. The acoustic impedance of mammalian long bone along, not transverse to, its long axis, has almost exactly the right acoustic impedance. Fabrication is tedious and unpleasant, however. A preferable solution is to use two matching layers, such that each is quarter-wave resonant and each lies at the geometric mean of the impedances of its neighbours. Thus if we take the acoustic impedance of PZT as $\sigma_1 = 30$ and that of water as $\sigma_4 = 1.5$ MRayl, the required layer impedances are

$$\sigma_2 = \sigma_1^{2/3}\sigma_4^{1/3} \qquad \sigma_3 = \sigma_1^{1/3}\sigma_4^{2/3}$$

or $\sigma_2 = 11$ and $\sigma_3 = 4$. Goll [7.10], who provides a comprehensive analysis of multiple matching layer transducers, suggests the use of glass and lucite to manufacture these layers. The author has developed other designs, more reliable in their manufacture, which work equally well. If carefully crafted, excellent octave-bandwidth transducers of high efficiency can be fabricated using this approach.

7.8 Polar Response Measurements on Transducers

The measurement of polar response is an important guide to transducer performance. A possible test-tank set-up is illustrated in figure 7.18. The equipment utilised may be as simple or as sophisticated as the user may wish or be able to afford. The most expensive item is usually the tank itself, together with its housing and installation cost. Ideally the tank dimensions should encompass many tens of wavelengths, if only because this would allow tone burst testing with range gating at the receive hydrophone to isolate the first received, direct path signal, before any reverberation (reflections from the tank walls and water surface) is received. The point here is, that most transducers have Q-factors of the order of 10 and thus take some ten cycles to complete their transient rise and fall times. If a useful dwell at maximum (i.e. a flat and measurable pulse top) is also required, then the need to be able to transmit perhaps fifty or so cycles per pulse becomes obvious. However, when it is recalled that one wavelength at 1.5 kHz is 1 m in extent, this is not always easily achievable in any but the largest tanks. Clearly the problem becomes less critical at higher frequencies. At frequencies of the order of hundreds of kHz, tank testing can be done in a water cistern in a laboratory at minimal installation cost.

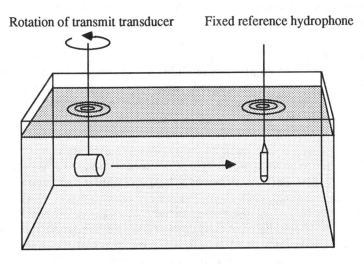

Figure 7.18. *Polar response measurement*

Tank lining materials have been proposed as a method of reducing or eliminating reflections from the walls, bottom and surface. For high-frequency work, in excess of 100 kHz, stipple rubber car mat works well. Rubberised horse-hair bats have also been suggested but the author has had little success with this material. At lower frequencies, marine ply baffle boards may be disposed about a large tank to break up the ordered nature of reflective returns. All such measures are, to an extent, cosmetic. Some value may be had from employing them or from searching for new approaches, but too much should not be hoped for. The moral of the story is: put your money in the tank; make it as large as you can afford in the first place.

The drive transducer will be rotated by some means, in order that the relative response at different angles of rotation may be measured by the fixed hydrophone. Rotation may simply be by hand, with a protractor on the rotating shaft used to measure angle. At the other extreme, a robust stepper motor drive, with microcomputer control of drive angle, transmitter timing and range gate timing may allow for automated measurement, data logging and display of polar response. In terms of today's technology, such a solution is neither difficult nor particularly costly.

7.9 Admittance Measurements of Terminal Response

The classic approach to measuring polar response, described well in Tucker and Gazey [7.11], is to plot the terminal admittance $Y = G + jB$ on the (G, B) plane. The author [7.12] has described a microcomputer-based system for making such measurements which avoids a commonly encountered axis-warping phenomenon experienced with some of the cheaper such equipments. However, the use of modulus of admittance $|Y(f)|$ plots is in some ways to be preferred, partly because it makes for an easier identification of certain key frequency and admittance modulus values required in the estimation of equivalent circuit values and partly because the experimenter may employ any convenient swept spectrum analyser to perform the measurement, provided only that it has available a swept sinewave output.

The easiest way to establish an equivalent circuit is that shown in figure 7.19. The tracking oscillator output of the spectrum analyser is used to drive the transducer, perhaps through a buffer amplifier although, being a substitution method, this should not really be necessary. Most spectrum analysers have a dual trace display with storage and if this facility is available, it should be used. The current into the transducer (proportional to the terminal admittance) is monitored by means of a current probe. A small-value sensing resistor may alternatively be used; the current probe will reflect into the line such resistance, anyway.

Acoustic Transduction

In performing the experiment, sweep frequency up from zero to some suitable frequency above the resonance frequency. Store the admittance trace so obtained. Remove the transducer and replace it with a capacitance box. Adjust the capacitance box value so that the straight line of $|Y(f)| = 2\pi C_0$ matches the slope of the stored transducer admittance at zero frequency, figure 7.19(b). In this way, the value of static capacitance is obtained. Next, figure 7.19(c), measure the resonance and anti-resonance frequencies, f_s and f_p respectively.

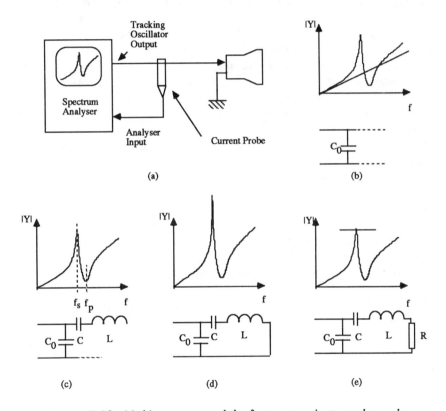

Figure 7.19. Making up a model of an acoustic transducer by measurement and substitution

Using the measured value of C_0 evaluate

$$C = (1 - (f_s/f_p)^2)C_0$$

and then evaluate the series inductance, L, by calculating

$$L = (4\pi^2 f_s^2 C)^{-1}$$

Again, using capacitance boxes and a tray of standard value inductors (or an inductance box if the reader is inclined towards self-indulgence in the fabrication of such an item; it can be well worth the trouble, for use in tuning experiments using a spectrum analyser, as well), set up the physical model, figure 7.19(d). Some "knob twiddling" may now be justified because the equations given only approximate the C and L values. In any event, connected as shown, a high-Q or "peaky" resonance should be obtained.

Lastly, add in a resistance box and adjust, figure 7.19(e), to reduce the Q-factor and make the synthesised and measured traces coincide, and the model is complete. This method will work, even for transducers with several significant resonances within the sweep band. In fact it is probably desirable to sweep upwards in frequency until no further significant resonances are identifiable, if the lowest resonance is not the one of interest, and then characterise the transducer in totality. Be warned, however, that any real transducer will exhibit myriad small resonances and clearly it will not be feasible to treat any but the most obvious. Some entertainment may be afforded by trying to establish the modal nature of the most significant resonances.

It is clearly possible, also, to improve the sophistication of the modelling operation. One may envisage an experiment which reads admittance data from the analyser, across to a microcomputer, which solves the equations and overplots the measured with a computed admittance curve, and so on. Coates and Maguire [7.13] have published equations for computing the equivalent circuit component values (to an approximation) for multimode transducers. They have also published the description of an iterative method [7.14] for refining the model so produced, so that the computed admittance response converges upon the measured response.

7.10 Hydrophones [7.15, 7.16]

In the main, our attention thus far has centred upon the projector transducer. Hydrophones, in some ways, offer less scope for variation of form to the designer and, because they may often be required calibrated, tend less to be the result of hand-crafting. Most hydrophones use PZT ceramic rings as their sensitive elements. Some utilise pairs of PZT half-spheres in an attempt to improve the omni-directionality of the device. A few utilise flat, or rolled, bar-mounted PVF film material. We shall focus attention on the first, largest class of devices.

Figure 7.20 shows the stages in the manufacture of a simple hydrophone. In this case, a single, capped tube element is soldered to the cable conductors

Acoustic Transduction 133

and encapsulated in a neoprene or polyurethane rubber. The design may utilise a tube of any appropriate size, subject primarily to the constraint that the hoop and length mode resonances occur at significantly higher frequencies than the hydrophone is intended to detect. The larger the diameter of the tube, the bigger will be its receiving sensitivity. It is also desirable that the tube wall be as thin as is practicable. Indeed, there are ratios of wall thickness to radius which can make the hydrophone virtually useless as a sound detector, because of a subtractive interaction between voltages generated within the crystal as a result of compression along orthogonal axes [7.15]. The hydrophone may also, with great advantage, be provided with an integral head amplifier, which will provide also a buffering and cable-driving action. If possible, the hydrophone should also be electrically shielded from external electric fields. This should be done by connecting the shield to the cable screen and thus to the surface amplifier input ground point. The shield should not be grounded through a sea-water return; this can prove extremely noisy. Some hydrophone designs provide also, a precision calibration resistor which is used for remote, field calibration purposes. Figure 7.21 illustrates a typical configuration. The hydrophone cable itself should exhibit low microphonics and great physical robustness. Twisted-pair conductors should be individually screened. For some applications, choice of a bitumen in-filling between the pairs may be

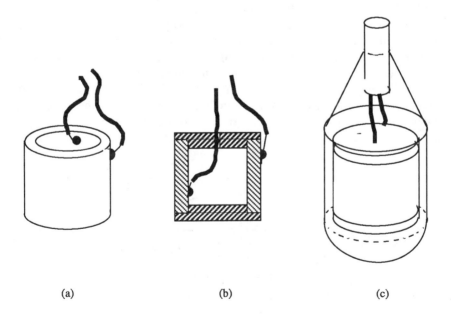

Figure 7.20. The basic hydrophone design using a PZT ceramic tube

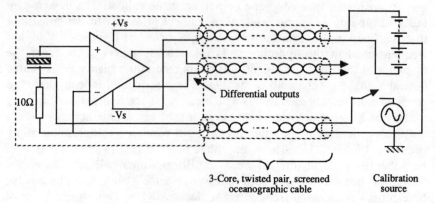

Figure 7.21 Hydrophone head amplifer, power supply and calibration circuitry

helpful in preventing the ingress of sea-water, should joints weaken or the sheath be damaged in any way. If a heavy hydrophone assembly with, perhaps, provision of a parabolic reflector and ballast weights is envisaged, then kevlar fibre reinforced oceanographic cable may be used.

Figure 7.22 shows the system outline for a hydrophone design used by the author for deep-ocean trials of a communication equipment. The underwater housing provided electronics which both passed a raw, but amplified received, modulated carrier signal to the surface, as well as a demodulated signal. During the course of the experiment, a completely separate equipment monitored and recorded both the raw signal and a demodulate, obtained in a different way. On board ship, two different receivers, as well as yet further recording capability were available. In fact, the direct microcomputer data input option shown in figure 7.22 worked perfectly but, obeying Murphy's Law, would certainly have failed had insufficient backup been provided.

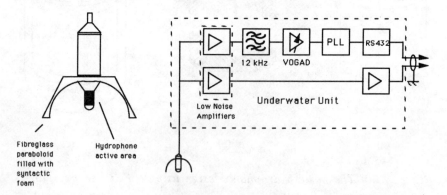

Figure 7.22. A data communications receiver hydrophone unit

References

[7.1] R.J. Urick, *Principles of Underwater Sound*, McGraw-Hill, New York, 1975

[7.2] P. Langevin, Brit. Pat. Specification, NS, 457, No. 145, 691, 1920

[7.3] J. Van Randeraat and R.E. Setterington, *Piezoelectric Ceramics*, Mullard Ltd, 1974 [ISBN 0 901232 75 0]

[7.4] L. Camp, *Underwater Acoustics*, Wiley, New York, 1970, p.136 ff.

[7.5] R.M. Davies, A Critical Study of the Hopkinson Pressure Bar, *Phil. Trans. Roy. Soc.*, Series A, Vol. 240, 1948, pp. 375-457

[7.6] D. Church and D. Pincock, Predicting the Electrical Equivalent of Piezoceramic Transducers for Small Acoustic Transmitters, *IEEE Trans. Sonics and Ultrasonics*, Vol. SU-32, No. 1, 1985, pp. 61-64

[7.7] R. Coates and R.F. Mathams, The Design of Matching Networks for Acoustic Transducers, *Ultrasonics*, Vol. 26, March 1988, pp. 59-64

[7.8] J.R. Dunn and B.V. Smith, Problems in the Realisation of Transducers with Octave Bandwidth, *Proc. Inst. Acoustics.*, Vol. 9, Pt 2, 1987, pp.58-69

[7.9] P. Wells, *Biomedical Ultrasound*, Academic Press, London, 1977

[7.10] J.H. Goll, The Design of Broadband Fluid Loaded Ultrasonic Transducers, *IEEE Trans. Sonics and Ultrasonics*, Vol. SU-26, No 6, 1979

[7.11] D.G. Tucker and B.K. Gazey, *Applied Underwater Acoustics*, Pergamon Press, London, 1966

[7.12] R. Coates, A Microcomputer-based Measurement System for Displaying the Admittance Characteristics of Acoustic Transducers, *Trans. Inst. Meas. Control*, Vol. 9, No. 4, 1987, pp. 218-223

[7.13] R. Coates and P.T. Maguire, Multiple Mode Acoustic Transducer Calculations, *IEEE Trans. Ultrasonics, Ferroelectrics and Frequency Control*, Vol. UFFC-36, No. 4, July 1988.

[7.14] R. Coates and P.T. Maguire, An Iterative Method for the Determination of Acoustic Transducer Lumped Equivalent Circuit Parameters, *Report No. SYS/E87/3*, School of Information Systems, University of East Anglia, Norwich NR4 7TJ

[7.15] O.B. Wilson, *An Introduction to the Theory and Design of Underwater Transducers*, Peninsula Publishing, Los Altos, Calif., 1988

[7.16] *BS5653: Specification for Hydrophones for Calibration Purposes*, British Standards Institution, London, 1978

8 Transducer Arrays

8.1 Introduction

In the previous chapter we examined the process of acoustic transduction and the way in which transducers might be assembled to meet particular design objectives. The common requirements in specification are:

1. Selection of transmission frequency
2. Selection of transmission bandwidth (or Q-factor)
3. Selection of power drive capability
4. Transducer electrical to acoustic conversion efficiency

In order to meet a given specification, additional consideration may need to be given to electrical termination and matching circuits. It may also prove possible to incorporate into the transducer design certain criteria regarding beamwidth and polar response pattern. However, in many situations it is preferable, or even essential, to separate these latter objectives from those of the design of the transducer elements themselves.

We should recall (section 7.1) that, for a circular aperture of diameter D transmitting sound of wavelength λ, the half-beamwidth of the polar response, measured in degrees, will be $30\lambda/D$. It is commonly the case in the design of a sonar system that beamwidth will, in fact, be a primary specification. Thus we might find it convenient to construct a piston transducer for, say, a 15 kHz ($\lambda = 10$ cm) sonar, by utilising a sandwich construction employing 50 mm diameter ceramic rings, an aluminium head-mass and a steel tail-mass. The overall diameter (and hence the effective aperture) could not be much different from 50 mm, which would suggest a full beamwidth of 120°.

If the initial design specification were for, for example, a full beamwidth of 20°, the implication would be that the transducer diameter should increase by a factor of six, to 30 cm. This is quite impractical, in a single, solid transducer design. The cost of ceramic material would be prohibitive.

Casting large ceramic pieces is extremely difficult and presents the manufacturer with a high failure rate and the customer with an extremely high cost product. The solution is to assemble an array of the smaller (~50 mm diameter) transducer elements (perhaps twenty or thirty in number) to provide the required aperture area. There are, needless to say, both advantages and disadvantages in such a strategy.

Among the advantages, we note that control of the relative "gain" of the peripheral elements in the array allows a measure of aperture "shading" to take place. This can help reduce sidelobe levels in the polar response, thereby minimising ambiguity in the discrimination between strongly reflecting off-main-lobe targets and weaker on-main-lobe targets. Set against this advantage, the array elements may well be prone to local interaction, making performance prediction difficult and introducing a costly "loop" into the array design process whilst such problems are ironed out. Yet a further advantage in utilising an array of transducers is that, by adding various electronic delay circuits to the individual transducers in the array, the polar response main axis may be "steered" electronically in any desired direction. This means that, to point the array in a given direction, we may dispense with servo-motor drives and mechanical interconnections to a soundhead and establish a virtually instantaneous shift of beam direction. It should be stressed, however, that the electronics required to achieve this is far more complex than that required for mechanical steering. The advantage lies primarily in speed of response: "inertia-less" beam-steering.

8.2 The Linear Hydrophone Array

The simplest array is a line of equally spaced, equally weighted (unshaded), omnidirectional elements. The elements are omnidirectional because they are dimensionally small by comparison with a wavelength of sound at the operating frequency. Such an array is illustrated in figure 8.1 in both an idealised form and a practical implementation. It would typically be used in marine seismic survey or in a military context, in the deployment of long arrays for passive, covert detection of enemy vessels. If it is of interest to detect frequencies of the order of 1.5 kHz or less (typical for both covert passive sonar and marine seismics) then wavelength will be of the order of 1 m or greater and thus much larger that typical array element dimensions. The array elements could well be capped ceramic tubes of one inch (2.5 cm) length and diameter.

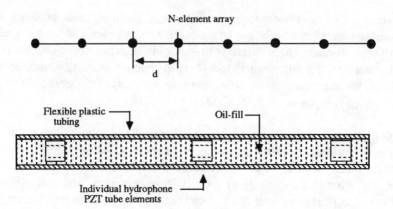

Figure 8.1. An N-element hydrophone array. Typically $d \sim 2$ m. $N \sim 24$ for seismic surveying; $N \sim 100 - 300$ for military arrays for passive detection, which could be up to 1 km in length if towed

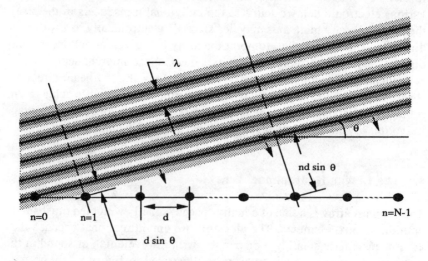

Figure 8.2. A plane parallel acoustic pressure field approaches a line array at inclination θ

Although the hydrophone array elements are presumed omnidirectional, the array itself will exhibit strongly directional properties normal to its direction of extension. To see why this is so, consider the array depicted in figure 8.2. A plane parallel wavefront, of wavelength λ, approaches at angle θ to the array. The relative spatial separation parallel to the direction of propagation, between the first and second array elements, is $d \sin\theta$. The relative spatial separation between the first and n'th elements is thus nd sinθ. Following the approach outlined in section 1.3, we represent the incoming perturbation as a travelling wave of unit amplitude $\cos(\omega t + kx)$. It follows that, for the n'th element, the signal output may be written in the form $\cos(\omega t + (n\omega d/c) \sin\theta)$. It is, at this point, manipulatively easier to proceed by noting that

$$\cos(\omega t + (2n\pi d/\lambda) \sin\theta) = \text{Re}\{\exp(j(\omega t + (2n\pi d/\lambda) \sin\theta))\}$$

$$= \text{Re}\{\exp(j(\omega t)) \exp(j(2n\pi d/\lambda) \sin\theta)\}$$

The array output we take to be the simple arithmetic sum of the outputs of the individual elements, scaled by a factor N^{-1} so that the polar response maximum is unity, irrespective of the number of elements in the array. It is possible, as we shall see later, to weight element outputs and, indeed, to introduce other forms of array processing than a simple arithmetic sum. Utilising this procedure, the array output becomes (the real part of)

$$\frac{\exp(j\omega t)}{N} \sum_{n=0}^{N-1} \exp(j(2n\pi d/\lambda \sin\theta))$$

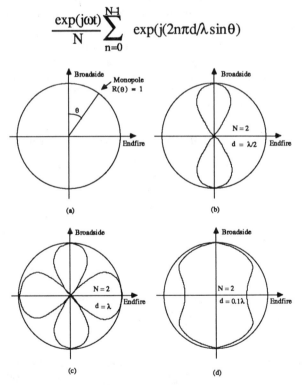

Figure 8.3. The response (a) of an omnidirectional monopole in the θ plane. Two monopoles in line form the simplest "interferometer" linear array, shown in (b) with half-wavelength spacing, in (c) with a full wavelength spacing and in (d) with 1/10 wavelength spacing

The time-dependence implied by the first term is of no interest to us. The summation in the second term may be shown to equate to the function

$$R(\theta) = \frac{\sin((N\pi d/\lambda)\sin\theta)}{N\sin((\pi d/\lambda)\sin\theta)}$$

which represents the directional sensitivity of the array. If $N = 1$, then $R(\theta) = 1$ for all θ; a single element is omnidirectional, figure 8.3(a). If $N = 2$, we obtain the simple two-element interferometer. It can further be shown, for this case, that $R(\theta) = \cos((\pi d/\lambda)\sin\theta)$. Given an element separation of one half-wavelength, we find that $R(\theta) = \cos((\pi/2)\sin\theta)$, which produces the directional pattern shown in figure 8.3(b), wherein a single ambiguity in direction identification occurs. We cannot know where, in the vertical plane (for a horizontally disposed array) a source will lie. If we increase the spacing to (say) one wavelength, the pattern becomes that shown in figure 8.3(c). Now the interferometer presents two ambiguities, in the sense that there are two principal axes along which a strongly detected (and thus presumably on-axis) source might be imagined to lie. Finally, if we reduce the distance between the elements so that they lie close together, relative to one wavelength then $R(\theta)$ again tends to unity for all θ, which is sensible, since two additively associated elements which are acoustically close, appear to merge as one, figure 8.3(d).

The reader should note that the two element interferometer is *not* the classical dipole (although a single element *is* a monopole). The dipole requires that the array element outputs are effectively differenced (either on reception or transmission, depending upon circumstance), not added. If we do this, we seek to establish a polar response

$$R(\theta) = (\exp(j\,\pi d/\lambda)\sin\theta - \exp(-j\,\pi d/\lambda)\sin\theta)/2$$

$$= j\sin((\pi d/\lambda)\sin\theta)$$

This equation we derive by imagining the dipole to be symmetrically disposed about the origin, so that the spatial displacements to each element are respectively $+(d/2)\sin\theta$ and $-(d/2)\sin\theta$.

The dipole is *presumed* to involve an element separation which is small by comparison with a wavelength. Then the response modulus becomes

$$|R(\theta)| = (\pi\delta/\lambda)\sin\theta$$

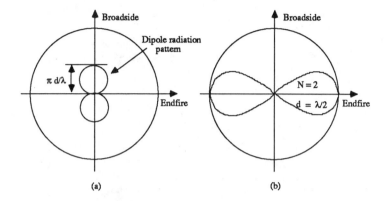

Figure 8.4. Polar response for the dipole element (small separation of elements) (a) and for the N = 2 array with 180° relative phase shift between the elements, forming a λ/2 separation end-fire array

which is the classical form of the dipole response already alluded to in describing sea-surface noise sources in section 6.5, and which is depicted in figure 8.4(a). Finally, we might enquire as to the effect on the polar response of accepting a wider spacing. If we make the element spacing on the (phase reversed) dipole elements one half-wavelength, the effect is to produce a 90° rotation on the polar response of the two-element interferometer pattern. We have essentially, by the implied phase shifting associated with the dipole geometry, converted a very short broadside array into an equally short endfire array, figure 8.4(b).

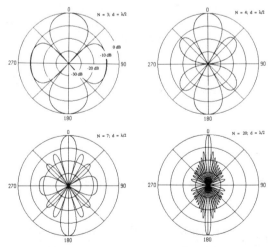

Figure 8.5. Polar response plots for unshaded, uniformly spaced arrays, with one half-wavelength interspacing between the elements. The "zero angle" direction equates to broadside sensitivity.

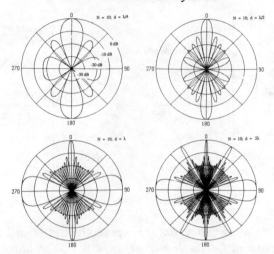

Figure 8.6. Polar response plots for unshaded, uniformly spaced 10-element arrays, with variable interspacing between the elements. The "zero angle" direction equates to broadside sensitivity.

Let us next investigate what happens as we increase the number of elements in our array. (We revert to the assumption, for the moment, of simple element output addition.) Figure 8.5 illustrates a progression through several stages indicating that as the number of elements is increased, for fixed d = $\lambda/2$ interspacing, the directivity also increases.

If, next, as figure 8.6 shows, we select a modest array length of 10 elements and then allow the spacing to vary, we see that for larger interspacing, multiple main lobes arise, because the array is effectively spatially undersampling the wavefield. For smaller interspacing, we worsen the directivity.

Finally, figure 8.7, we see the effect of fixing the array length, Nd = L, to be constant, whilst causing d to approach zero. The bottom line in our expression

$$R(\theta) = \frac{\sin((N\pi d/\lambda)\sin\theta)}{N\sin((\pi d/\lambda)\sin\theta)}$$

then tends to $N\pi d/\lambda \sin\theta$ and the polar response assumes the familiar "sinc" or "sin(x)/x" form:

$$R(\theta) = \frac{\sin((N\pi d/\lambda)\sin\theta)}{(N\pi d/\lambda)\sin\theta} = \frac{\sin((\pi L/\lambda)\sin\theta)}{(\pi L/\lambda)\sin\theta}$$

This pattern corresponds to the uniform, linear array with even spacing across its aperture.

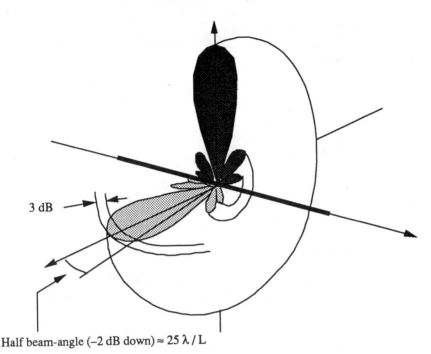

Half beam-angle (−2 dB down) ≈ 25 λ / L

Figure 8.7. Directional response of a continuous line array of length L

8.3 The Fourier Transform Approach to Pattern Synthesis

In the previous section, in summing the contributions from individual elements from a discrete array, we anticipated a more formal approach to the determination of a far-field radiation pattern from the way in which a hypothetical acoustic aperture was filled. In the most general sense, we may assume the existence of some two-dimensional aperture, a function of x and y, say. This aperture will produce a far-field response which will be a function of the normalised sines of the beam angles in the horizontal and vertical: $\Theta = (2\pi/\lambda)\sin\theta$ and $\Psi = (2\pi/\lambda)\sin\psi$. The inter-relation is the two-dimensional complex Fourier Transform

$$W(x,y) \Leftrightarrow R(\Theta,\Psi)$$

The reader is referred elsewhere [8.1] for proofs. Here we note the one-dimensional simplification

$$R(\Theta) = \int_{-\infty}^{+\infty} W(x)\exp(-j\Theta x)\,dx$$

and, by way of example, reconsider the uniform line array referred to at the end of the previous section. If we define the array to be of length L, uniformly illuminated, so that

$$W(x) = L^{-1}; \quad |x| < L/2$$
$$= 0 \text{ elsewhere}$$

then the Fourier Transform $R(\Theta)$ follows as a standard result from any suitable Table of Transform pairs, as

$$R(\Theta) = \frac{\sin(\Theta L/2)}{\Theta L/2}$$

Upon making the necessary substitutions $L = Nd$ and $\Theta = (2\pi/\lambda) \sin\theta$, our previous result will be found to have been obtained. The radiation pattern for the continuous line array has already been illustrated in figure 8.7. Inspection of the pattern reveals that it has fallen to $-3dB$ of its main axis value for an angular off-axis shift (the nominal "half beamwidth") of $25\lambda/L$ degrees.

The method is a powerful one for investigating the behaviour of far-field response. We may, for example, apply "shading" to the array aperture. In the example referred to above, the one-dimensional aperture, the continuous line array was unshaded. Equal weighting was given to all increments along

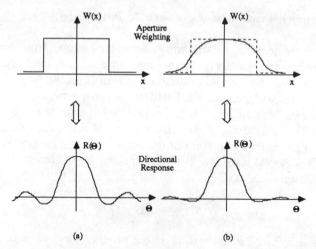

Figure 8.8. The use of aperture shading to reduce the sidelobe level of an aperture (a) with uniform illumination. To maintain the same main lobe width, the shaded aperture (b) should be physically wider, but with illumination tapering off at its extremes

its length. Had we so chosen, we could have applied any of a large family of aperture weighting functions to modify the sensitivity, in the case of a receiving array, or the power output, in the case of a transmitting array. In particular, if we choose to smooth the transition at the end of the array aperture, then, as figure 8.8 suggests, we may reduce the size of the sidelobes in the radiation pattern and thus reduce the hazard of false target identification, without significantly losing angular resolution. That is, if the sidelobes are prominent, there is a danger that an off-axis target of relatively large target strength may be mistaken for an on-axis, albeit weaker target. Not surprisingly, perhaps, the mathematics of weighting closely parallels the windowing applied in numerical power spectrum evaluation. The reader is referred to the common source of information on this topic [8.2] and to [8.3], where several examples pertinent to aperture shading are quoted.

The method may also be applied to provide the evaluation of certain notable results. The first of these pertains to the uniformly illuminated circular aperture. This is equivalent, in effect, to a single circularly symmetric acoustic projector and the result obtained for the far-field response is thus of considerable importance as a first-cut design aid. The reader might care to note that, the transducer being axially symmetric, the Fourier Transform may be modified to yield the result

$$R(\Theta) = 2 \frac{J_1(\Theta L/2)}{\Theta L/2}$$

where Θ has the same meaning as before, L is the diameter of the transducer and J_1 is the first-order Bessel function "of the first kind". The radiation pattern is loosely of "sinc" form, with an axially central main lobe nested within a cone of sidelobes of decreasing amplitude, figure 8.9. In this case the nominal half-beamwidth is $30\lambda/L$ degrees.

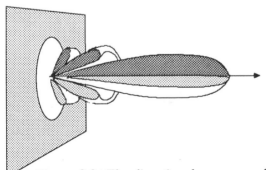

Figure 8.9. The directional response of a circular aperture in an infinite baffle, approximating to the response of, for example, an air-backed circular ("piston") transducer element.

8.4 Array Beamsteering [8.4–8.6]

The simple hydrophone array illustrated in figure 8.2 involved only the summation of all hydrophone outputs, to "form" the beam. No shading was used, nor was there any other form of processing following reception. The result, for $\lambda/2$ spacing and a reasonable number of elements within the array, was an angularly selective main-lobe directed in the broad-side direction, normal to the direction of extension of the array itself. In examining the dipole, however, we noted that, with just two elements if, instead of simply adding outputs, we inverted (phase shifted by 180°) the output of one of the hydrophones before adding, then the broadside pattern would twist through 90° and become an endfire pattern.

This principle may be extended for the general array to produce a beam-steering effect which allows us to direct the narrow main lobe at any angle we wish, to the line of extension of the array itself. Thus consider the array implementation shown in figure 8.10. Here, the n'th element has, imposed upon its output, a delay $n\tau$. The effect will be to produce a beam-swing through an angle θ given by

$$\theta = \sin^{-1}(c\tau/d)$$

Of course, array shading may also be applied, in order to tailor the main-lobe shape and minimise sidelobes. Indeed, quite complicated array processing may be envisaged, which can allow the creation of steerable nulls in the polar response, which can be used to eliminate a return from an unwanted, off-axis target. Arrays constructed on this principle may take on a wide

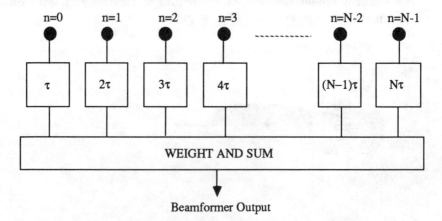

Figure 8.10. The "inertia-less" electronically steered beamformer array

Transducer Arrays

range of physical forms. Line and plane arrays have obvious benefits in allowing transmitter or receiver beams to be formed and directed. However cylindrical and part-spherical arrays are also used. It should also be mentioned that, simple as the beamforming concept may appear, there are many practical difficulties to be overcome in the development of effective beamformers. In particular, the delay synthesis electronics is complex and costly and there will almost inevitably be problems with element-to-element interaction in the array itself.

8.5 Directivity Index

One consequence of improving the directional properties of a sound source by forming an array is that the available power is directed within a smaller solid angle than would have been the case if an omni-directional source had been used. In this context, we have already seen, in section 2.3, that if the array subtends a solid angle ϕ, then the on-axis acoustic intensity is raised by a factor $4\pi/\phi$. Expressed in decibels this factor is the directivity index, DI.

The quantity ϕ is calculated, if the array radiation pattern $R(\theta,\psi)$ is known, by evaluating the integral

$$\phi = \int_0^{2\pi} \int_{-\pi/2}^{+\pi/2} R(\theta,\psi) \cos\theta \, d\theta \, d\psi$$

If the radiation pattern has rotational symmetry, the integral may be further simplified thus

$$\phi = 2\pi \int_{-\pi/2}^{+\pi/2} R(\theta) \cos\theta \, d\theta$$

Applying the result to our expressions for the continuous line of length L, we find that $\phi = 2L/\lambda$, and to the circular aperture of diameter L, we find that $\phi = (\pi L/\lambda)^2$.

8.6 The Parametric Source [8.7–8.9]

In all our considerations thus far, in this chapter, we have dealt solely with arrays of transducer elements so combined as to produce given transmission characteristics. In the main, we find that desirable characteristics are: high sensitivity combined with directional angular response and freedom from excessive sidelobes. In this and the next section, we investigate two other methods of achieving high angular resolution, but with "non-physical" apertures of large spatial extent. It should be stressed that, although both methods to be discussed utilise transducers which are physically small by comparison with the transmission wavelength, none the less the aperture remains large, as indeed it must if effective beamwidth is to be small.

The first approach to creating a large non-physical aperture using a relatively small transducer is known as "parametric" transmission. If a low-frequency transducer is driven at adequately high power, a phenomenon known as "cavitation" will be seen to occur. Cavitation is manifest as bubble formation in the water in front of the transducer, coupled with a pronounced fluid streaming, away from it. The effect occurs because, at high power drive levels, the pressure fluctuations in the water can become so extreme as literally to draw gas out of solution, tearing the water apart, as pressure falls below ambient. Cavitation becomes more difficult to induce at depth, because yet higher pressure amplitude fluctuations are needed. It is also more difficult to induce if frequency is increased. Because cavitation represents, in a sense, the approaching of an upper limit to possible pressure amplitude fluctuations, studies of sound generation at such levels are referred to as "finite amplitude acoustics". Fully developed cavitation, by and large, is to be avoided in underwater acoustics because it can lead to erosion damage on the face of high-power transducers. The effect finds practical application in another context, in ultrasonic cleaning equipments.

The propagation of sound in water relies on the fact that the magnitude of pressure fluctuations is directly related to the magnitude of particle velocity fluctuations. At the levels of pressure fluctuation most frequently encountered in underwater acoustics, the relationship between these quantities is linear and described by the equation, analogous to Ohm's law in electrical circuit theory

$$p = \sigma u$$

Here p is rms pressure fluctuation, u is rms particle velocity and $\sigma = \rho c$ is, of course, specific acoustic impedance. If transducer face vibrations become sufficiently great, then as cavitation is approached, this relationship breaks down and the fluid medium begins to behave in a non-linear fashion.

If, then, we cause to be launched into the water by a high-power projector, two frequency tones f_1 and f_2, these tones will, as it were, spatially co-exist in the water column in front of the projector. Since the water behaves non-linearly, a power-series development is possible, wherein a multiplicative mixing may be anticipated, producing along the length of the column in front of the transducer, products of the form $\cos(\omega_1 t)\cos(\omega_2 t)$. These products may be re-written as sum and difference frequencies $\{\cos(\omega_1 t - \omega_2 t) + \cos(\omega_1 t + \omega_2 t)\}/2$. It is only the difference frequency that is of interest to us. Although the non-linearity produces only fairly weak mixing and hence difference frequency generation at low power levels, relative to those of the primary frequencies, the physical sites of difference frequency generation are spatially distributed in front of the projector for a considerable distance. This means that a spatially large, effectively endfire aperture at the difference frequency can be obtained. This in turn implies high directivity at the difference frequency.

In practice, we should normally choose primary frequencies in the region of hundreds of kHz. Then we should have available some tens of kHz of transducer bandwidth and an ability to synthesise a difference frequency of up to half the projector bandwidth. This means that a very wide, albeit relatively low-frequency, sweep of frequency is possible by the parametrically generated secondary. Also, because the sidelobes at the two primary frequencies do not exhibit any particular spatial correlation, there is virtually no sidelobe structure at the difference frequency.

Parametric generation thus provides us with a method of generating low-frequency sound, with extreme frequency agility, high angular resolution and almost total freedom from sidelobes. The penalty is low efficiency of conversion: in the region 1 to 10%. It has been suggested that an acoustically transparent tube, containing fluid of lower vapour pressure than water, be placed in front of the projector, in order to enhance the parametric generation. The reader is referred to a paper by Muir [8.9] for further information on the practical design of parametric sonars. The reader is also warned that difference frequency generation can also occur within the drive power amplifier, within any matching transformer or tuning inductor used with the projector transducer, within the transducer itself and, indeed, within a receive hydrophone or its associated electronic circuitry. This is because, at high levels of drive, all these equipments may be expected to offer significantly non-linear behaviour in some respect. Great care is needed, therefore, in ensuring that the (weak) generation mechanism being sought really is parametric in nature.

8.7 Synthetic Aperture Sonar [8.10–8.13]

If parametric sonar provides a virtual endfire array, then synthetic aperture sonar establishes a virtual broadside array. The synthetic aperture principle is widely used in terrain mapping radar [8.4]. It has, for example, been used

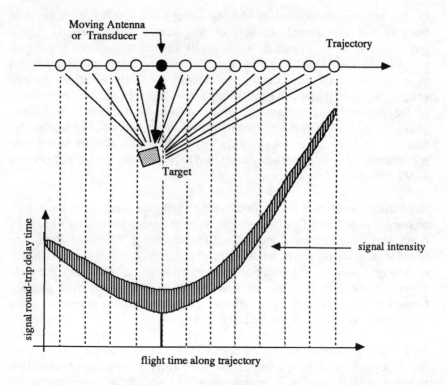

Figure 8.11. Non-coherent synthetic aperture sonar

for radar mapping both from aircraft and from earth and planetary reconnaissance satellites. A major requirement with airborne or spaceborne radars is that the antennae should not be overly large, for purely practical aerodynamic reasons. This militates, at radio frequencies, against achieving angular resolution which adequately complements range resolution capability. The concept underlying the synthetic aperture principle is that, if an antenna can be physically translated through space, it can act successively as the individual elements on a long, linear array. The way in which this can occur is illustrated in figure 8.11 in the context of "non-coherent" aperture synthesis.

The non-coherent synthetic aperture is realised by recognising that the crescent-shaped pattern of received signals on the graph of signal round-trip delay time versus flight time of the antenna is a characteristic identifier of the target. Indeed, cusp-shaped features are a commonplace on the output of the dry-paper recorders such as are used in conjunction with many modern echo-sounders. Whenever the echo-sounder records a strong but small-scale feature such as a rock-scaur or a dense school of fish, the crescent form may be noticed. Of course, the echo-sounder does not seek to make use of such information. However, it is not difficult to imagine that some numerical method could be devised to gather up the spatially distributed information in the crescent and present it as a high-resolution point at the target location. In fact, some airborne synthetic aperture radars utilise a film recording process, whereby specially designed conical lenses re-focus such features. Although it is relatively easy to appreciate the nature of the synthetic aperture principle by considering the non-coherent case, it is actually preferable to measure both the amplitude and the phase of the incoming signals. Then the delay time versus flight time diagram becomes, effectively, a zone-plate hologram and the reconstruction is, essentially, holographic in nature, whether performed optically or numerically.

For many years, the possibility of creating a practical synthetic aperture sonar has been held in question. Three objections to the possible efficacy of the technique are commonly quoted. First, it was supposed that the stability and control of any towfish could not be accurate enough to maintain an adequately linear trajectory. The required trajectory should hold to a straight line, to within a fraction of a wavelength, over many complete synthesised apertures. The author recalls, for example, a photograph published in the mid 1970's which purported to show an actively controlled US Navy experimental towfish for synthetic aperture research. The nose-cone bore the legend "Murphy's Law" attesting, we may presume, to the many experimental difficulties encountered with the device. However, depending upon context, there seems no reason why the problem of towing stability should be insuperable, or that the tow geometry and trajectory should not be sufficiently accurately monitored to make suitable correction possible. For example, there is some current interest in the use of synthetic aperture to enable low-frequency, long towed arrays to be reduced in size. Then wavelengths are such as to suggest that towing errors would be handleable.

A second objection had to do with water path inhomogeneity. Again, whilst this might be a problem in some localities or with some particular applications, such as high-frequency, ultra-high resolution sidescan employed in estuarine site survey situations, the difficulty does not, at lower frequencies and in open ocean water, appear to be serious. The third reason had to do with the fact that, because sound travels so much more slowly that light, the sonar – unlike the radar – would exhibit an unacceptable spatial aliasing problem. The use of frequency modulation encoded transmission appears to circumvent this last difficulty [8.13]. Whereas it is reasonably certain that the synthetic aperture principle can be made to work for active sonar, in terrain mapping and other applications, it is less certain that the technique is applicable to passive target detection. Finally, it is interesting to note that, whereas in a conventional sonar, a large physical aperture is essential for good azimuth resolution, with a synthetic aperture, the moving antenna or transducer, the physical aperture should be small. This is because the desirable mode of operation is omnidirectional insonification of the field being mapped.

References

[8.1] R.N. Bracewell, *The Fourier Transform and its Applications*, 2nd edition, revised, McGraw-Hill, New York, 1986

[8.2] F.J. Harris, On the Use of Windows for Harmonic Analysis with the Discrete Fourier Transform, *Proc. IEEE*, Vol. 66, Jan. 1978, pp. 51-83

[8.3] C.S. Clay and H. Medwin, *Acoustical Oceanography*, Wiley, New York, 1977, pp. 138-177

[8.4] M.L. Skolnik, *Introduction to Radar Systems*, McGraw-Hill, New York, 1980

[8.5] Special Issue on Beam Forming, *IEEE. J. Oceanic Engineering*, Vol. OE-10, No. 3, July 1985

[8.6] Special Issue on Underwater Signal Processing, *IEEE. J. Oceanic Engineering*, Vol. OE-12, No. 1, January 1987

[8.7] H.O. Berktay, Possible Exploitation of Non-Linear Acoustics in Underwater Transmitting Applications, *J. Sound Vibration*, Vol. 2, p. 435, 1965

[8.8] H.O. Berktay, Some Finite Amplitude Effects in Underwater Acoustics, in *Underwater Acoustics*, Vol. 2 (V.M. Albers, ed.), Plenum Press, New York, 1967, pp. 243-261

[8.9] T.G. Muir, Non-linear Acoustics and its Role in the Sedimentary Geophysics of the Sea, *Physics of Sound in Marine Sediments* (L.L. Hampton, ed.), Plenum Press, New York, 1974, pp. 241-287

[8.10] L.J. Cutrona, Comparison of Sonar System Performance Achievable Using Synthetic Aperture Techniques with the Performance Achievable by More Conventional Means, *J. Acoust. Soc. Am.*, Vol. 58, No. 2, 1975, pp. 336-348

[8.11] L.J. Cutrona, Additional Characteristics of Synthetic Aperture Sonar Systems and a Further Comparison with Non-synthetic Aperture Sonar Systems, *J. Acoust. Soc. Am.*, Vol. 61, No. 5, 1977, pp. 1213-1217

[8.12] P. de Heering, Alternate Schemes in Synthetic Aperture Sonar Processing, *IEEE J. Oceanic Engineering*, Vol. OE-9, No. 4, October 1984, pp. 277-280

[8.13] P.T. Gough, A Synthetic Aperture Sonar System Capable of Operating at High Speed and in Turbulent Media, *IEEE J. Oceanic Engineering*, Vol. OE-11, No. 2, April 1986, pp. 333-339

9 Sonar Engineering and Applications

9.1 Introduction

In this chapter we examine the way in which electronics and acoustics interact to provide practical solutions to a wide range of sonar engineering problems. It is a matter of some regret that the history attaching to the development of the subject of underwater acoustics is relatively poorly documented. As history, the documentary material is of relatively recent origin and is frequently difficult to gain access to because of the explicitly military nature of much of the research which has been conducted during the past several decades. An excellent general account, of a largely non-technical nature, has been published by Haines [9.1]. This text views the development of underwater acoustics from a British standpoint but with copious reference to contributions made elsewhere. Urick's introduction [9.2] provides a brief historical perspective, which is nicely complemented by the first background chapter in a recent text by Burdic [9.3]. A text edited by Albers [9.4] provides a collection of benchmark papers of particular interest to the underwater acoustician.

Reference is frequently made to what is possibly the earliest written reference to the detection of sound at sea, which was made by Leonardo da Vinci [9.5], writing in 1490: "If you cause your ship to stop, and place the head of a long tube in the water and place the outer extremity to your ear, you will hear ships at a great distance from you." Despite the obvious mismatches in acoustic impedance involved in using such a principle, variants of this technique (almost certainly without regard to Leonardo's original observation) were employed during World War I, both at shore listening stations and on surface vessels, for the primary purpose of submarine detection and localisation. The simplest such device, which can easily be replicated and tested, consisted of a pair of submerged air-filled rubber bulbs separated by about two metres, connected by stiff-walled tubes to a pair of stethoscope earpieces. The device has directional characteristics for frequencies in the medium and high audible range.

The first electrical sensors utilised carbon button microphones contained in flexible rubber tubing some 10 metres in length, to form a towed array. This device was introduced during the last months of World War I and was a development which specifically recognised the need to remove the sensor from the self-noise generated by the ship deploying it. Again, simple microphonic detectors based upon this principle are easily made. Suitable button microphone elements are still to be had in army surplus stores: they form a part of the throat-microphones used by aviators in past years.

Although the early objective uses of sonar were certainly of a military nature, it was the loss of the Titanic in 1912 that first stimulated thought into ways of detecting objects in the sea – in this instance, icebergs rather than submarines – by using active ultrasonics. In 1914, Fessenden was able to demonstrate the detection of an iceberg at two miles range using a moving-coil transducer [9.6]. Work by Langevin, Chilowsky, Boyle and others lead to the development, in the years during and following World War I, of quartz transducers, and metal–quartz–metal sandwich transducers of the type discussed in section 7.3. From this time but particularly during the period of World War II, with the rapid development of electronics as a valuable complementary technology, sonar engineering became firmly established as a scientific and engineering discipline in its own right.

In the period following World War II, commercial developments connected with marine civil engineering survey, hydrographic survey, oil-prospecting, scientific investigations and sub-sea site operations, such as navigation and re-location, have taken underwater acoustics out of its narrower military framework, making it a major enabling technology for a wide range of maritime activities. This burgeoning in application has gone hand in hand with vital developments in transducer materials, in particular the discovery of the lead titanates and their generic successors. Equally important has been the rapid increase in complexity and decrease in cost of electronic components and systems for signal generation and processing and, indeed, the dramatic progress in development of appropriate algorithmic techniques for a host of signal processing activities.

9.2 The Basic Echo Sounder

The simplest and most widely used of all sonar equipments is the Echo Sounder, which is to be found on all Merchant and Military shipping and, in any of a number of economical designs, on even the most modest of small pleasure and sporting craft. In essence the device is quite simple, consisting of a transmitter amplifier capable of generating a voltage pulse of sufficient amplitude to launch, almost always by means of a piezo-electric transducer, an acoustic sounding pulse vertically downwards into the water. Often, although not exclusively, the same transducer will act to pick up the received signal, which will be suitably amplified and then detected to allow estimation of the travel time to the sea-floor and back. In pursuing the following description, emphasis will be placed on the "unclever" parts of the circuitry. The microprocessor-based, real-time signal processing is not usually the greatest problem. Being sure that the electrical engineering is reliable and the mechanical design well thought out requires great care.

Figure 9.1 provides a block diagram of the basic echo sounder. Operation is as follows. The master clock generator issues timing pulses whose period is equivalent to the round-trip delay for a pulse reflected from the sea-floor at maximum depth. Thus, for an instrument required to operate in the range 0–150 m, the maximum round trip is 300 m and the pulse repetition period is 200 ms. Each master clock pulse initiates the generation of a fixed-width, pulsed sinusoid. For a small-boat echo-sounder intended to operate in shallow water with reasonably good depth resolution, choice of operating frequency is often dominated by the availability of cheap 150 kHz transducer elements. If greater depth capability is required, a lower frequency – some few tens of kHz – will be used.

The RF pulse waveform is then amplified. The amplifier will be designed to generate peak power at a level appropriate for a particular application. A power of the order of watts will be required for the high-frequency, small-boat echo sounder, tens or even hundreds of watts for low-frequency deep

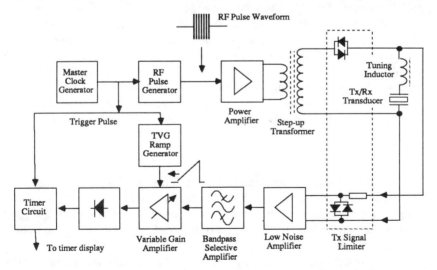

Figure 9.1. The basic echo sounder shown in schematic form

water echo sounders. The objectives in designing a suitable semiconductor power amplifier are: low output impedance, high efficiency and adequately high current output capability. The last of these criteria is important because with semiconductor circuitry, the power line voltages will be very much less (some tens of volts) than the drive voltages (hundreds or even thousands of volts) which will typically be needed to generate the design power levels from the (relatively high impedance) transducer. Consequently, the power amplifier will be required to supply high current to the primary of a transformer which will have a substantial turns ratio.

The transformer itself will be required to be quite carefully designed. At the drive frequencies envisaged for virtually all echo-sounder applications (although not necessarily for sub-bottom profilers, which operate at frequencies of the order of 1–5 kHz) ungapped ferrite cored assemblies will be mandatory. For the higher frequency applications, these may be of the pot-core type. Otherwise, E-core assemblies will probably be found to be more appropriate. If the transformer is properly designed, the core *magnetic* volume will be relatively unimportant. Core size will most probably be dictated by the volume of copper needed to accommodate primary current with safe turn-to-turn voltage difference and by the fact that winding layers in the secondary will require to be individually insulated, to cope with the need to place hundreds or even thousands of volts across the transducer element.

It is, of course, possible to wind the transformer and measure, using a suitable meter or bridge, the secondary inductance. This inductance cannot be used to tune the static capacitance of the transducer crystal. To see why the transformer, if restricted to provide a high-efficiency voltage step-up, should not be used to provide tuning, it is necessary to recall that the high coupling coefficient of an ungapped ferrite core results in extremely low leakage-flux and thus only very small values of primary and secondary *leakage* inductance L_p and L_s. Note that leakage inductance is not the same as the primary and secondary *winding* inductance. The winding inductances *should* be large. Indeed, they should be infinite for an ideal transformer. Furthermore, they will also be effectively unseen by any terminating circuitry.

Any secondary tuning must derive from the leakage inductances, referred to the secondary, where they will appear in series with the referred amplifier output impedance. The amplifier output impedance should, with good design, be negligible. Even the total referred leakage inductance for a transformer with a 1:N turns ratio, which will be $L_s + N^2 L_p$, will usually be far too small to provide tuning. Of course the leakage inductance could, in principle, be increased to a value which might tune the transducer. This would, however, require air-gapping of the core. This, in turn, would increase the core reluctance, to the detriment of effective power transfer. The inevitable conclusion is that separation of function is advisable: let the transformer act primarily as an impedance changer, and employ a series inductor, specifically designed for the purpose, to cancel static capacitance.

Design of the inductor is not necessarily straightforward. In many respects, it presents a significantly more difficult task than designing the output transformer. The following point should be borne in mind. If the tuning inductor begins to saturate, it will act as a non-linear circuit element. In the following account the reader will be reminded of describing function theory, in control engineering. Suppose, first, the drive voltage to be modest, with the inductor behaving as it should, without evidence of core saturation. Suppose also the drive frequency, the tuning inductance value and the static capacitance to be such as to engender resonance. Now gradually increase the drive voltage. The voltage across the inductor will increase proportionally. So also will the current through it, and its core flux. At some point, the flux will rise to such a magnitude that saturation will begin to occur. A clipping of the current waveform will result. The saturating inductor will then effectively enter a phase shift into its current waveform which, since resonance is critically phase sensitive, will immediately de-tune the resonant circuit with consequent loss of power.

There is a rather crude moral in this story. If the transducer you wish to tune, hefts heavy in the hand, do not expect to tune it with a 15 mm pot core! A considerable volume of ferrite will be required to hold adequate energy within its magnetic field, per half cycle, to pass on to the electro-mechanical resonator which is the transducer.

At this point, the amplified RF pulse has been delivered to the transducer, from which it is emitted as a high-power acoustic pulse. The pulse travels to the sea-floor, reflects with some loss, and returns to the same transducer, which acts reciprocally to detect its presence and convert it into an electrical signal. It is true that two transducers can be used, one to transmit and one to receive. This approach increases cost in some measure. In some sonars, it is imperative to use separate transmit and receive transducers, if different insonification and inspection characteristics are required. This is often the case with terrain-mapping sector-scanning sonars, for example. However, it is when a single transmit/receive transducer is used that circuit design complications result, so that is the situation we treat here. Clearly, because of the extremely high drive voltages during transmission, it is inappropriate merely to connect the receive amplifier directly to the transmit/receive transducer. Instead a solid-state changeover network must be employed. One such is shown in figure 9.1. During transmission, the parallel diodes in series with the transformer secondary present a low impedance connection between transformer and transducer. On reception, when the voltage across the transducer will be small, perhaps no more than millivolts in amplitude, the diodes will present a relatively high slope impedance, effectively isolating the transmission circuitry from the transducer. Again, during transmission the shunt diodes across the amplifier input will conduct, presenting a fraction of the transmit signal no more than the forward diode knee voltage in amplitude, to the receiver. On reception, again because of the relatively high slope resistance for input signal amplitudes with a magnitude much less than the knee voltage, the shunt diodes will appear as an open circuit. More complex, active switches can be devised but the circuit shown in figure 9.1 will work well in many applications.

The first receive amplifier will almost always be chosen to be a solid-state low-noise design, using either discrete components or an ac coupled, integrated low-noise amplifier. Although the transducer will be to some extent selective, it is most probable that it will respond to a number of unwanted frequency bands, because of the presence of spurious resonances. It may well be necessary to make the first low-noise amplifier at least broadband selective at the transmit/receive frequency. If this is not done, it is possible that out-of-band ambient noise may swamp the amplifier, driving it into saturation and thus inhibiting the detection of the wanted signal, despite any subsequent bandpass filtering.

Making the front-end amplifier broadband selective will not normally provide enough bandpass filtering to eliminate all the out-of-band ambient and front-end receiver noise. It is customary to introduce several stages of selective, narrowband bandpass filtering prior to detection, in order to achieve this.

The fact that the acoustic signal both spreads and is attenuated in passing to the sea-floor and back, means that the received signal amplitude falls, roughly as the square of water depth. In order to compensate for this effect, a time variable gain (TVG) circuit is usually incorporated. This is simply an amplifier, or cascade of amplifiers, whose gain is dynamically increased by a ramp control voltage waveform triggered by the transmit pulse.

Finally, envelope detection will allow the received RF pulse reflected from the sea-floor to be isolated and prepared for use as the round trip timing pulse. The timing circuitry is relatively straightforward and in a modern echo sounder might present its output on a numeric display, or on a monitor screen in any of a variety of formats, or on some form of dry-paper chart recorder capable of presenting a "facsimile" display.

9.3 Sub-bottom Profiling

As has been mentioned, the basic echo sounder finds use in ship and small-boat navigation where, more often than not, its function is that of a warning device, to indicate inadequate under-keel clearance. It also finds use, in much the form described here, but with operation at only some few kHz, in sub-bottom profiling. The use of these lower acoustic frequencies reduces the potential resolution but allows penetration into marine sediments, permitting geotechnical inspection of the sea-floor, as well as possible location of buried objects such as wreck artifact, cables or oil-pipelines. Figure 9.2 shows a sub-bottom profiler output. Features beneath the sea-floor, such as sediment layers and rock outcrops, can be clearly identified.

Because high power is required to penetrate the sediment, a range of alternatives to the more conventional piston-type piezo-electric transducer (section 7.3) has been devised. These include sparkers, where a bank of high-value capacitors is discharged across a spark-gap under water, and a range of air-gun and explosive detonation devices, synchronised in some way to the master clock. Repetition rates for such applications are likely to reflect, respectively, the spatial scale of the sub-bottom features being

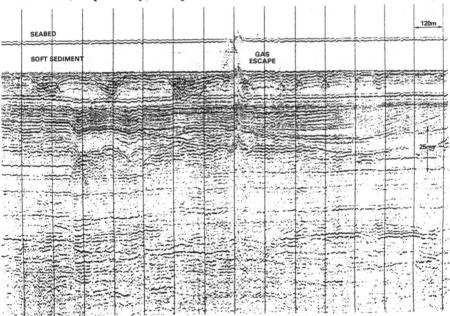

Figure 9.2. A sub-bottom profiler record, taken off the West African Coast, showing a sediment layer above consolidated sediments representing the channels of an ancient river delta. A gas emission can also be seen on this record (courtesy of Gardline Surveys Ltd, Great Yarmouth)

investigated, and the survey speed. Because the acoustic aperture of reasonably easily handleable low-frequency soundhead arrays will be relatively small, and the beam-spread and resolution uncomfortably large, this is an area of application where parametric sonar (see section 8.6) may yet prove valuable. Yet further future improvements in sub-bottom profiling sonars are likely to involve spread-spectrum encoding of the transmitted waveform to assist in improving range resolution.

9.4 Fishing Sonars [9.7]

The simplest echo sounders can be used, with skill, in fish-finding applications, since a clear return signal is often obtainable from fish swimming in shoals. However, the application is one of commercial importance and a range of improvements upon the basic echo sounder have been made to enhance its value to the fishing community. Clearly, a "facsimile" type display, such as produces a strip, dry-paper record of the water column beneath a boat, as the boat progresses on it way, is of value as a diagnostic aid since, in addition to providing a permanent record of past events, it allows a measure of visual integration, making possible the detection of features which, on a scan-to-scan basis, might not be apparent.

One shortcoming of the dry-paper record is that little information concerning received signal strength is available. In order to improve interpretation yet further, the detected output of a basic high-frequency ($\sim$100–200 kHz) echo sounder may be processed to yield a gray-scale or, even better, a false-colour display of received signal strength as a function of time. Yet a further sophistication is to be found in the use of several possible different transmission frequencies. The idea behind multi-frequency echo-sounding is that the reflective characteristics of different fish sizes, species or sea-floor types at different frequencies, might in some sense provide improved diagnostic information. Ideally, one might wish for an echo sounder capable of emitting an ultra broad-band sounding pulse, extending from some tens of kHz to the high hundreds of kHz. The acoustic returns from such a pulse could then, effectively, be spectrum-analysed within the epoch containing energy reflected from a particular feature of interest, such as a fish-shoal. The major problem in devising such an equipment lies in the design of adequately broadband, efficient and cheap acoustic transducers with a suitably uniform frequency response. Currently, multifrequency sonars use but a restricted set of narrowband transmissions and are capable of only modest manipulation of the data so acquired.

Most fishing echo sounders operate in a vertical transmission mode. Some are, however, adaptable to provide a rotating sweep at some reasonably acute angle to the horizontal. This makes possible a plan-position indication of sub-sea features, which could be either on the sea-floor or in mid-water. The method is thus particularly attractive for the remote location of shoals of fish or other commercially valuable shoaling marine life such as shrimp or squid. Whilst it is clearly simplest to rotate the scanning soundhead mechanically, there has been much interest in either totally electronic rotational scanning, or partially electronic and partially mechanical scanning. The technique used is referred to as sector scanning sonar and involves the design of arrays and beam-forming electronics such as was described in section 8.5. A major obstacle to the wider adoption of sector scanning sonar is the high cost of the multi-element transducer arrays and beam-forming electronics, and the rapid escalation of this cost with demand for increased cell resolution.

In the context of horizontally-directed plan position indicating sonars for fisheries research, interesting work has been conducted on the development of miniature transponding tags which can be sewn without harm to the skin of a fish to enable fish-tracking experiments to be conducted. The tags, an example of which is shown in figure 9.3, are activated by the transmitted

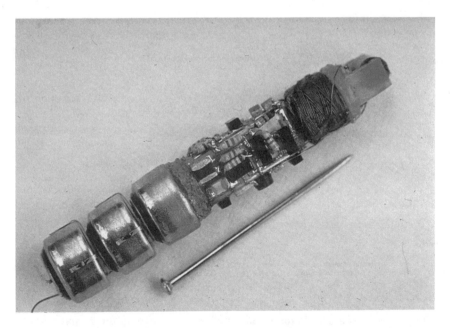

Figure 9.3 Miniature transponding fish tags (courtesy of Fisheries Laboratory, Ministry of Agriculture, Fisheries and Food, Lowestoft)

sonar beam, and return a signal at the same frequency, but with a much greater signal strength than would the fish being tracked. So sophisticated has become the development of these transponding tags that measurement and re-transmission of compass direction data, to a resolution equal to "eight points of the compass rose", has been achieved [9.8].

Yet another aspect of fishing sonar development has to do with quantifying the size of a marine population, either by counting or by some attempt at biomass estimation. The former approach involves the use of narrow-beam, high-frequency echo sounders of high spatial resolution. The latter involves monitoring the acoustic backscatter from a population, when insonified by a directional high-frequency echo sounder, and using this signal to infer the biomass density [9.9].

9.5 Side-scan Terrain-mapping Sonars

The side-scan sonar utilises much the same electronic system architecture as the basic echo sounder. It provides an alternative solution to the plan-position indication problem referred to in the previous section. Side-scan

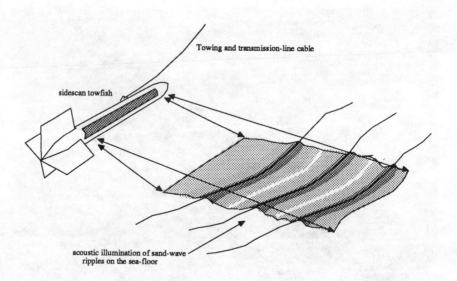

Figure 9.4 The side-scan sonar towfish, typically about a metre long and with a narrowband pulsed transmission at a centre frequency in the band 100–500 kHz

(a)

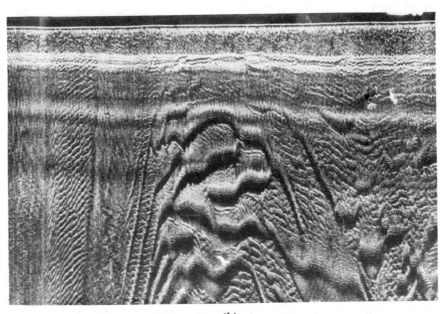

(b)

Figure 9.5 Side-scan sonar images: (a) a wreck in the North Sea, showing deck detail and, thrown into the sand, the pale acoustic shadow, showing details of superstructure (courtesy of Gardline Surveys, Great Yarmouth); (b) sand water ripples on the sea-floor (courtesy of Fisheries Laboratory, Ministry of Agriculture, Fisheries and Food, Lowestoft)

sonar is widely used in marine site- and pipeline-surveying, and is important in mine countermeasures, in locating and identifying enemy mines. The principal features of the side-scan system are illustrated in figure 9.4. The most important feature of the equipment is the soundhead itself which, being many (typically about 50) wavelengths long, is of unusually narrow beamwidth in the horizontal plane, and thus provides good azimuth resolution.

Whilst it might be possible to mount such a soundhead on a surface vessel, much as the ordinary echo sounder soundhead is mounted, but set in a sideways looking attitude, several factors militate against this. First, servo stabilisation of the soundhead would be needed, to counter the effects of pitch and roll while the vessel was under way. This is entirely feasible, and is done for some hydrographic survey echo sounders of a specialised nature. However, it is a technique which adds substantially to cost. Utilising the towfish principle effectively eliminates the stabilisation problem.

It also removes the soundhead from the self-noise of the towing vessel. Since echo returns will be at oblique incidence and therefore rather weak, this is at least beneficial. Furthermore, the soundhead can be towed nearer the sea-floor than if ship-mounted, and this improves the short-range swath coverage. Finally, the entire equipment can be transported easily and can be rapidly set up on even quite small vessels of convenience.

The side-scan sonar will typically operate to generate about 100 W of pulsed acoustic power at a frequency in the range 100–500 kHz. The swath width will typically extend to about 100 m on either side of the towfish, which will carry identical soundheads on each side. Each outgoing pulse will illuminate a lateral patch on the sea-floor and the return signal from this patch will establish one scan-line of a "raster scan" picture of the sea-floor, which builds up as an image on, usually, a dry-paper facsimile-type recorder, as the vessel proceeds along its tow. As the image shown in figure 9.5 shows, considerable detail of the sea-floor may be obtained. Sandwave ripples, rock-scaurs, exposed pipeline, cable and anchor chain, as well as wreck and other debris, may be clearly identified by a well-executed survey. The higher frequency sidescans will offer the best resolution and in the band 100–500 kHz there will be relatively little degradation of range. The author is aware of sidescan designs in the low MHz range, with a swath width of the order of 25 m, which have been investigated for mine identification purposes.

At the other extreme of operational scale, "GLORIA", the Geological Long Range Inclined Asdic, developed at the United Kingdom Institute of Oceanographic Sciences, and currently operated under contract by Marconi Underwater Systems Ltd, utilises a sidescan array which is about 5 m long by 1.5 m high, operating at a transmission frequency of 6.5 kHz with a pulse length of 30 ms and a pulse transmit power of 50 kW. It can cover a swath of width up to 27 km to one side of the towfish and, as its name suggests, was originally designed for rapid sea-floor mapping activities for geological investigations [9.10].

9.6 Seismic Survey [9.11]

In this sphere of activity we move farthest from the structure of the conventional echo sounder, although the operating principle remains broadly the same. We have seen how, by using transmission frequencies of a few kHz, penetration of the sea-bed may take place and sediment layering and other features may be revealed. For geological studies and, particularly, for offshore gas and oil prospecting, dramatically deeper penetration of the sea-bed is needed, often to depths in excess of a kilometre. In order to achieve such penetration, two major changes in system design are necessary. First, it becomes mandatory to use low-frequency (1 kHz and below) "seismic" shock waves, rather than piezo-electrically generated acoustic pulses. These

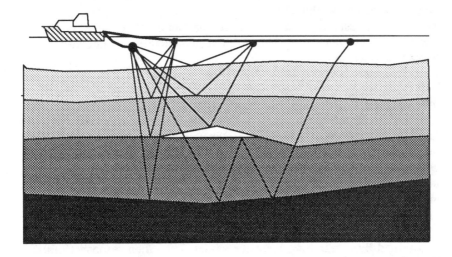

Figure 9.6. *The survey ship detonates a shock wave (short tow cable) which passes on many paths to (three of many) hydrophone elements on the long towed array. Sediment acoustic properties (sound speed, density) vary within the layers. A salt-dome enclosing a gas pocket is also shown*

shock waves are typically generated either by an air-gun, which releases compressed air rapidly into the water, producing a bubble field very similar to that released by a dynamite detonation, or by an oxygen–propane gas-detonation in a flexible-walled chamber within a shallow-towed submerged raft. Second, it is necessary to use a long towed array as the receiver. The array will typically contain some two to four dozen hydrophone elements within a flexible, oil-filled plastic tube of diameter (about) 5 cm and length up to 3 km.

The shock wave generated on each firing by the survey ship passes downwards, reflecting off the sea-floor to produce, at the various hydrophones in the array, a first reflected shock wave. Further reflected shock waves will result from interaction with the sediment interfaces. Particularly strong reflections will be identified from gas pockets beneath impermeable salt-domes, because of the marked change in acoustic impedance at such an interface.

The data processing activity which takes place in analysing the time sequences resulting from successions of shock-wave detonations as the ship executes a survey is one of the classical "inverse problems" of mathematics. Time of flight on many reflective and refractive passes through many sediment layers will be known, but layer thickness, sound speed and density will not. The task is to establish, and refine the parameters of, a model predicting these unknown quantities. This is done by numerical calculation, with human interaction.

9.7 Acoustic Positioning and Navigation [9.12]

In discussing the use of fish-tags for tracking purposes, in section 9.4 above, the concept of the transponder was introduced. Transponders are equipments which, deployed in pairs or in larger networks, can interact with each other, to allow determination of their separation by acoustic pulse time-of-flight measurement. The simplest transponders are acoustic beacons of the type used in conjunction with aircraft flight recorders. Such beacons remain in a passive, listen-only mode, until awoken by an interrogation pulse, usually transmitted by a searching recovery vessel. They then respond to further interrogation pulses allowing, to some degree, the surface vessel to position itself vertically above the beacon, at which time the transponding round trip delay will be a minimum.

The method is extended in underwater navigation systems, by emplacing a net of at least three slave transponder units at particular geographic locations. The master transponder is able to activate coded responses from each of the beacons so that, as figure 9.7 illustrates, slant ranges between

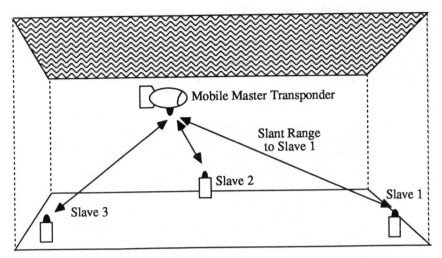

Figure 9.7. The circular acoustic navigation principle

the master and each of the slaves may be measured and, by a process of triangulation, location within the survey area determined. The literature of the subject distinguishes between long-baseline and short-baseline systems. This distinction is less than completely fundamental. It does not (generally) relate the terms "long" and "short" to, for example, transmission wavelength. Rather, it tends to reserve the former terminology for sea-floor slaves and the latter for ship- or platform-mounted equipments.

The method illustrated in figure 9.7 is unambiguous spherical navigation. The mobile transponder may be thought to be at the unique point of intersection of three hemispheres centred on each of the three slaves and of radii equal to each of the respective slant ranges. The intersection of only two such hemispheres would produce a vertical semicircular locus of possible position, rather than a uniquely determined point of fix.

Circular, or transponding navigation carries with it the penalty that the mobile must carry an active transmitter. In radionavigation applications, this would prove to be a severe disadvantage, with many users competing to communicate with the slave stations. The problem can be avoided, and the active transponder replaced by a passive receiver, if hyperbolic rather than spherical navigation is used. Consider for example, the geometry illustrated in figure 9.8. Here, two beacons and the mobile are shown located within a single inclined plane surface. On this surface are shown the hyperbolic loci which mark out lines of constant time difference in reception of signals which had been emitted *simultaneously* from the two beacons. If the mobile measures what it "hears" as the time difference, it can determine on which hyperbola it lies. Naturally, with only two beacons the system is ambiguous. The ambiguity may be entirely or contextually removed by overlaying one or more further patterns of hyperbolae and/or utilising additional information, such as depth data. Hyperbolic navigation in underwater survey is less frequently employed than spherical navigation.

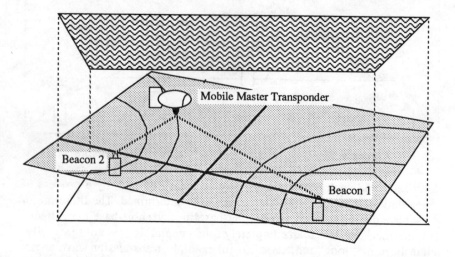

Figure 9.8. Hyperbolic acoustic navigation

9.8 Doppler Measurements

The doppler effect is the frequency shift which occurs in perceived sound when either an observer moves with respect to the transport medium, or the medium itself moves. The effect is employed to advantage in several sonar measurement equipments among which is the doppler current meter. This device, illustrated in figure 9.9(a), is used to measure water velocity at a point, or movement of an object through the water. The operating principle is that of a high-frequency continuous wave sonar with spatially separated transmit and receive transducers. The horizontal flow component along the sound axis between heads A and B and heads C and D of the meter generates doppler shifts of magnitude

$$\Delta f_{AB} = f_0(v/c)\cos\theta \quad \text{and} \quad \Delta f_{CD} = f_0(v/c)\sin\theta$$

From these equations, speed v and the direction angle θ are easily obtained.

A loosely similar principle is used in doppler logs. The Janus configuration, illustrated in figure 9.9(b), is frequently used in this application. It allows the forward speed and sideways drift of a submersible to be determined. The calculations involved are similar, except that an additional angular dependence is introduced by the downward depression angle of the beams. The doppler shift information is contained in backscattered, rather than reflected, sound.

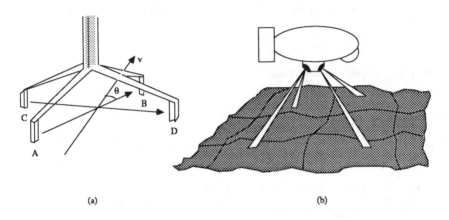

Figure 9.9. *Applications of doppler measurement in underwater acoustics*

A range-gated doppler log has recently been successfully operated for ocean remote current sensing. Here the concept is to inspect the doppler shift on signals backscattered in the consecutive range cells of a pulsed high-frequency sonar. The instrument takes advantage of the signal processing power available using modern microelectronics.

Finally, the doppler principle has been employed with considerable success in monitoring sea-floor geotechnical properties [9.13]. Here, a constant frequency 12 kHz transmitter is housed in the tail of a free-fall, torpedo-shaped projectile. As the projectile, which is known as a penetrator, descends it accelerates, with corresponding doppler shift, until it reaches a terminal velocity which, for a two-tonne penetrator can exceed 100 miles per hour (50 m s^{-1}). At this speed the received signal at the surface is approximately 11.6 kHz. On impact, the penetrator decelerates and the doppler shift decreases to zero. This allows the deceleration profile to be measured and the depth of penetration to be calculated. This in turn allows the sediment strength to be estimated remotely, without coring.

References

[9.1] G. Haines, *Sound Underwater*, David & Charles, Newton Abbott, 1974

[9.2] R.J. Urick, *Underwater Sound for Engineers*, McGraw-Hill, New York, 1975

[9.3] W.S. Burdic, *Underwater Acoustic System Analysis*, Prentice-Hall, New Jersey, 1984

[9.4] V.M. Albers, *Underwater Sound*, Dowden Hutchinson and Ross, Stroudsberg, Penn., 1972

[9.5] E. MacCurdy, *The Notebooks of Leonardo da Vinci*, Garden City Publishing Co., New York, 1942

[9.6] R.A. Fessenden, US Patent Application 744,793 (1913)

[9.7] R.B. Mitson, *Fisheries Sonar*, Fishing News Books, Farnham, 1983

[9.8] N.D. Pearson and T.J. Storeton-West, The Design of an Acoustic Transponding Compass Tag for Free-Swimming Fish, *Proc. 5th IERE Intl. Conf. on Electronics for Ocean Technology, Edinburgh, September 1987*, pp. 83-92

[9.9] R. Coates and C. Orgill, Population Density Measurement by Acoustic Backscatter, *IEEE Oceans '87 Conference, Halifax, Nova Scotia, September 1987*

[9.10] R.B. Whitmarsh and A.S. Laughton, A Long-range Sonar Study of the Mid-Atlantic Ridge Crest near 37°N (FAMOUS Area) and its Tectonic Implications, *Deep Sea Research*, Vol. 23, 1976, pp. 1005-1023

[9.11] E.A. Robinson and T.S. Durrani, *Geophysical Signal Processing*, Prentice-Hall, London, 1985

[9.12] P.H. Milne, *Underwater Acoustic Positioning Systems*, Spon, London, 1983

[9.13] R. Coates, *A Deep-ocean Penetrator Telemetry System, IEEE J. Oceanic Engineering*, Vol. 13, No. 2, April 1988, pp. 55-63

10 Acoustic Communications

10.1 Introduction

The literature surrounding the subject of underwater acoustic communications is, in some respects, surprisingly scant. This is particularly the case if one concentrates only upon that material directly concerned with actual underwater communication systems as opposed to more general aspects of propagation and channel modelling. The major difficulties with which the communication engineer is concerned, when attempting to design underwater acoustic communication systems, revolve around the problems of reverberation and multipath transmission and high attenuation at high acoustic frequencies. An extensive bibliography on the subject has been published by the author [10.1]. A subset of that bibliography, dealing with selected specific underwater communication systems is presented here [10.2–10.13]. However, the reader is referred to the original source for more information on specific communication systems, and papers on general aspects of channel modelling and more detailed mathematical treatments than can be handled in a text of this nature.

Acoustics is not, of course, the only method of obtaining underwater communication: cable systems (in many different contexts) are widely used – but carry the obvious and frequently unacceptable disadvantage of tethering the remote end of a link. Fibre-optic methods have also been used, to a lesser extent, as yet – though doubtless that situation will change in the future. The problem of tethering remains, as does the need for copper, in providing a supply of electrical power to the sub-sea installation.

Electromagnetic propagation, using both radio and laser transmission, has been considered. However, except in particular and unusual circumstances, neither approach is of great value. Because salt water is conductive, only the lowest – barely usable – radio wavelengths will propagate any distance. These are emanations in the ELF (Extra Low Frequency – 30 Hz to 300 Hz) band. Signals propagated in the ELF band require large transmitter powers and large antennae. ELF band propagation has been studied intensively during the past two decades with a view to establishing "bellringer" communication with the nuclear submarine fleet. That is, given a channel of such restricted information bandwidth, simply use it to request, from a normally covert submarine fleet, use of surface-deployed high-frequency radio antennae, to download a rapid stream of short-term, tactically valuable information.

Curiously, visual light frequencies are the least attenuated of all electromagnetic emanations, by salt water. The idea has even been proposed – the reader must judge the significance of the proposition – that the eye evolved as it did because it evolved in primitive sea-creatures, for which the only significant stimuli for nerve-cell evolution, when the animals themselves were submerged in the sea, were in the optical waveband. Unfortunately, from the communication engineer's viewpoint, the difficulty with optical communications lies less with attenuation than with scattering. It is the presence of the numerous scattering particles in the sea which militates against long-range optical communication. The use of laser light, because of its narrow, pencil beam, certainly serves to minimise the scattering volume between transmitter and receiver, but the technique remains fraught with difficulties and, at this time, is largely impracticable except in very specialised application areas.

10.2 The Gross Attributes of the Received Signal

If, as figure 10.1 suggests, the received waveform consists of the sum of a main path, plus reverberation paths, then certainly the received signal will exhibit some degree of fading behaviour. This we may classify in terms of fading statistics which will describe the probability density distribution of envelope and phase respectively.

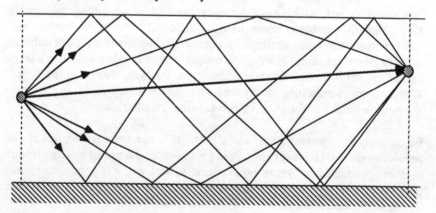

Figure 10.1. Multipath structure in a highly reverberant environment

If the main path is not grossly the dominant path, then the classical model for such phenomena, based upon the application of the central limit theorem, will presume Rayleigh amplitude and uniform phase distributions. That is

$$p(A) = (A/<A^2>) \exp(-A^2/2 <A^2>); \quad 0 \leq A \leq \infty$$
$$= 0; \quad -\infty \leq A \leq 0$$

where $<A^2>$ is the variance, or mean square value, of the envelope $A(t)$.

$$p(\phi) = 1/2\pi; \quad -\pi \leq \phi \leq +\pi$$

If the main path is in fact dominant, then the amplitude distribution may tend towards being of Rician [10.14] form.

Another attribute which may be of value in gauging the effectiveness of possible solutions to the multipath problem will be the spectral bandwidth of the fading envelope. Here, we attack the problem by considering an heuristic view of the processes giving rise to the envelope fluctuations themselves. Consider the situation in which a main-path and a single, dominant, grazing incidence surface bounce occurs. Suppose surface wave action is slight: just a slow, small amplitude shift of the reflective surface of the sea. Then we may anticipate long, slow and deep fades whenever the multipath and main received signals move into phase opposition. However, the frequency with which these fades will occur will be greater than the surface wave frequency, if the length difference between the direct path and the reflected path spans many cycles at the carrier frequency. Thus given range, R, depth of receiver and transmitter, h, and assuming an average surface wave height σ, the differential length will be

$$\delta l \approx 4h\,\sigma/R$$

If both transmitter and receiver are firmly located in the water, then it is interesting to consider the frequency of envelope fading caused by a single surface reflection interference path acted upon by surface waves. The geometry of the problem is illustrated in figure 10.2.

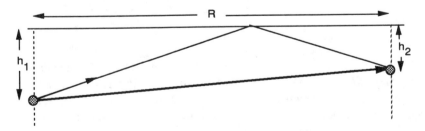

Figure 10.2. *Transmission dominated by a single surface bounce and the main path*

The number of cycles of fade per surface wave cycle will be $\delta l/\lambda$ where λ is the transmission wavelength. Both the average height σ and the average frequency f_{avg} of surface wave cycles are in turn dependent upon wind speed, w [m s^{-1}], being given approximately by the relations

$$\sigma = 5 \times 10^{-3} \times w^{2.5} \qquad f_{avg} \approx 2w^{-1}$$

and it follows that the number of cycles of fade per second or, loosely, the fading frequency of the envelope, will be of the order of

$$\frac{0.04hw^{1.5}}{R\lambda}$$

Thus if, for example, we transmit at 15 kHz, for which $\lambda = 0.1$ m, over a range of 500 m, at a depth of 10 m and with a modest wind speed of 20 m s^{-1}, we find that the fading frequency will be about 0.7 Hz.

Another interesting condition relates to the vertical spatial frequency of interference maxima, since this quantity gives at least some indication of the vertical interspacing of receiver array elements, if some form of space diversity is contemplated as a method of combating the fading problem. Here, we assume the transmitter to be at depth h_1 and the receiver to be at depth h_2. If the differential length is equal to some integer number of wavelengths $\delta l = n\lambda/2$; n = 0,1, ... then a condition for constructive interference, incorporating the 180° phase shift at the sea-surface, will be met, as figure 10.3 shows, at receiver depths of the order of $h_2 \approx n\lambda R/4h_1$.

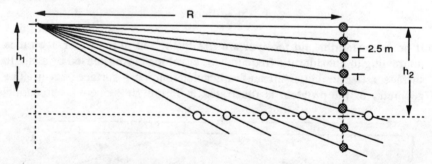

Figure 10.3. The fan lines are lines of destructive interference in the R–z plane. Shaded blobs correspond to vertically defined intensity nulls at range R. Unshaded blobs correspond to horizontally defined intensity nulls at depth h_2 (the moving ROV situation)

Acoustic Communications

Let us again consider the example of a 15 kHz source at 10 m depth and 500 m range. The vertical interspacing of intensity nulls will then be 2.5 m, suggesting array element interspacing of this order, if space diversity reception is contemplated.

Yet a third condition relates to the frequency of envelope fading if either the transmitter or receiver is moving; mounted, for example, on an autonomous ROV. Then, we find that the fading frequency for a receiver at depth h_2 moving towards or away from a transmitter at depth h_1 and at range R will be approximately

$$v(4h_1h_2 + R\lambda)/R^2\lambda$$

which for our 15 kHz source, with both transmitter and receiver at 10 m depth, and with their horizontal separation being 500 m, results in a fading frequency of 0.03 Hz if the ROV speed is 4 knots, or 2 m s^{-1}.

In conclusion, these simplified examples of operation may lead us to suppose that the required response bandwidth of electronic equipments designed to combat fading need only be modest. For our example, the fastest fading frequency noted above, to surface wave induced fading, was only 0.7 Hz. For safety, and bearing in mind the simplicity of the model, we might imagine, given other conditions, that assuming a fastest fading frequency in the regime 10–100 Hz should allow the system designer some latitude in formulating schemes to combat multipath.

10.3 The Channel Transfer Function

Let us take, again by way of example, the situation depicted in figure 10.3. The differential path length $\delta l \approx 2h_1h_2/R$ imparts a differential delay $\delta t = \delta l/c$. Note also that the two paths will be subject to attenuation caused by spreading and loss but that, for our present example both loss factors will be comparable. There will be a surface reflection loss, k, but at grazing incidence and modest wind speed, this will also be modest. The channel impulse response will thus be of the form

$$h(t) + k\,h(t + \delta t)$$

Here $h(t)$ is the main path transfer function, dictated primarily by the transmission and reception equipments. The loss factor is given by $k = -10^{\mu/20}$ where μ is the intensity reflection coefficient and where the negative sign models the 180° phase shift on reflection from the sea-surface. Fourier transforming, we find that the channel frequency response is

$$H(jw) + k \exp(j\omega\delta t) H(j\omega) = A(\omega) \exp(j\phi(\omega))$$

where

$$A(\omega) = |H(j\omega)| \{2(1 + k \cos(\omega\delta t))\}^{1/2}$$

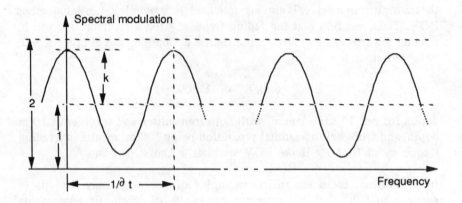

Figure 10.4. Spectral modulation resulting from the interference between a surface bounce and a main path

For our example (R = 500, h_1 = 10, h_2 = 10, λ = 0.1) we find that δt = 267 μs. It follows, as figure 10.4 shows, that the transfer function $|H(\omega)|$ has imposed upon it, in forming $A(\omega)$, a frequency ripple. For the stated channel properties, this ripple has a frequency period of some 3.745 kHz.

If the (15 kHz) transmit transducers exhibit a nominal (−3 dB) bandwidth of 1.5 kHz then, if k is unity, intersymbol interference of uncertain but possibly considerable severity may be anticipated. Thus constructive interference (a maximum of the spectral modulation) at the transmit centre frequency will result in the relatively unaffected transmission characteristic illustrated in figure 10.5, whereas destructive interference (a minimum of the spectral modulation) at that frequency will engender severe spectral distortion and serious intersymbol interference.

For our example, a higher transmission frequency and thus wider bandwidth will produce a spectral modulation which may introduce several cycles of frequency variability in the transmission characteristic. If the chosen values of R, h_1 and h_2 were altered, so that δt increased, then again we should expect faster amplitude fluctuations across the transmission characteristic $A(\omega)$. Finally, with yet further multipaths entered into our model, higher order anharmonic ripples in the transmission characteristic will result, leading to a complex modification to the mainpath transfer function, itself fluctuating at a rate not dissimilar, we might presume, to the fading frequency discussed above, and presenting some significant problems in adaptive equalisation.

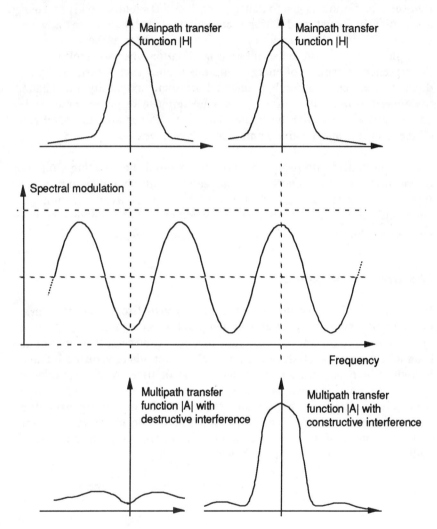

Figure 10.5. *The effect of a single surface bounce plus the main path, in bringing about spectral distortion on the channel transmission characteristic*

10.4 Combating Multipath

The simplest approach to overcoming multipath, in whatever context the transmission occurs, is just to transmit the data sufficiently slowly and using a sufficiently robust modulation technique that the channel has adequate time to settle and allow detection decisions to be made at the receiver. For example, if we employ pulse interval modulation, then we seek to detect the presence of packets of energy consequent upon each transmitted pulse, distinct from each other but smeared to some extent by the channel reverberation time. It is thus the channel impulse response, or effective bandwidth, which limits the transmission rate, rather than the bandwidth of the transmit and receive equipments themselves.

Clearly, both diversity reception, with its potential for selecting situations of constructive interference, and equalisation, with its ability to minimise intersymbol interference and maximise the use of available mainpath bandwidth, are attractive propositions whether invoked independently or simultaneously.

10.5 Diversity Reception

In the context of underwater acoustics we may visualise three broad classes of diversity: spatial, spectral and temporal diversity. Spatial diversity systems take cognisance of the fact that the multipath structure of the channel can, indeed, produce regions in the water where acoustic intensity is high, albeit only for some limited period of time. A simple diversity system, such as is shown in figure 10.6, may thus be employed to allow the receiver to select the hydrophone element producing the highest output. Such a system implies no beamforming, is simple to implement and thus offers the merit of some degree of element redundancy, in the event of hydrophone failure, and thus enhanced reliability.

Naturally, such a system may be modified, as figure 10.7 suggests, to allow far more sophisticated use of the available hydrophone elements. The function of the weighting networks is to provide gain and phase adjustment on the signals received from each hydrophone. Here a preferred technique is maximal ratio combining [10.15] which attempts to maximise received signal-to-noise ratio. Maximal ratio combining carries with it a substantial computational penalty which may be outweighed for underwater applications by the simplicity of either the switched diversity receiver or an equal gain combining scheme. It should be noted that, in a sense, the law of diminishing returns applies. If switched diversity picks out the strongest signal, relatively small benefit accrues from adding in several other signals of increasingly dubious quality.

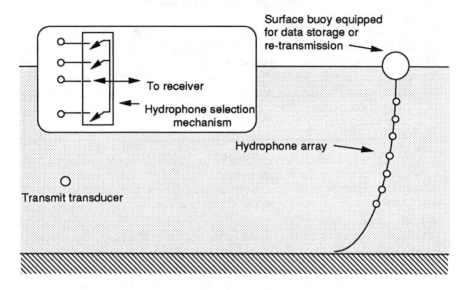

Figure 10.6. Simple switched (spatial) diversity reception

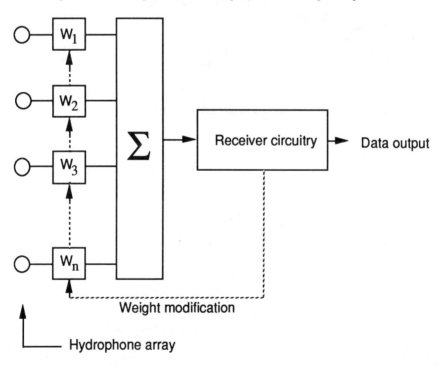

Figure 10.7. Maximal ratio combining in linear diversity detection

It is possible, of course, to imagine the weighting as a beamforming operation. It would appear that maximal ratio combining has the effect of steering nulls of the array polar response into the direction of the multipath "image" sources. One would imagine, in any case, that the beamforming weightings would have to account for the somewhat random amplitude and phase factors experienced in the sound-field as a consequence of the multipath environment.

Spectral diversity takes account of the fact that, because of the varying multipath structure of the channel, so also will the channel transfer function vary as a function of time. If transmission may be made, either switchable or simultaneously, on more than one frequency, then thus may we anticipate or utilise a momentarily preferable frequency slot for signalling. Just as beamforming "looks a bit like" some forms of spatial diversity, so frequency-hopped spread spectrum communications resemble, in some respects, spectral diversity signalling. Some care should be exercised in making such intuitive jumps, if for no other reason than that transmitter and receiver transducers may be of distinctly limited bandwidth. We return to this topic, in the context of parametric transmission, in the last section.

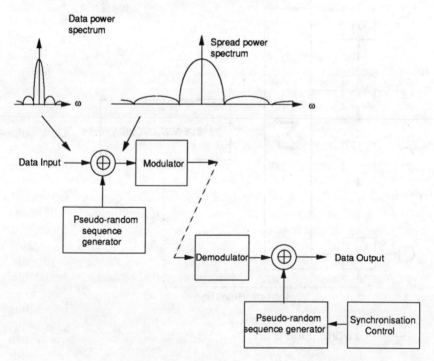

Figure 10.8. Direct sequence spread spectrum

Temporal diversity acknowledges the transport-delay model of the channel. If coding can be applied such that information is spread over a time-frame significantly longer than the channel reverberation time, then it is possible that error corrective properties may be utilised in overcoming the multipath problem. Again, an analogy can be drawn with direct sequence spread spectrum communications used, in this instance, specifically to reject the multipath interfering signals. Figure 10.8 illustrates the basic system configuration.

A data source with bit-rate R_1 is added, modulo-2, to the output of a pseudo-random digit generator, with a digit or "chip" rate, R_2. Normally, $R_2 \gg R_1$, so that the spectral bandwidth of the PRDG output would be much greater than that of the data sequence. The digital "modulation" thus impressed on the data sequence can be removed at the receiver by adding exactly the same PRDG sequence. Synchronism is thus a vital presumption. The nature of the synchronism must be such as to account for the main-path transport delay. If the multipaths add in yet further delay, the signal returns arriving by those routes will not synchronise and will be suppressed in the ratio $(R_1 : R_2)$, in terms of their relative power, by the receiver modulo-2 addition operation. Of course, for this suppression to be unambiguous, there is an implication that the sequence length $T = L/R_2$ must be at least greater than the longest significant additional multipath delay. If it is not, then there is the possibility that a multipath return may occur at a time equal to some multiple of the sequence length and be decoded as though it were a valid main-path return.

10.6 Equalisation

Equalisation is the process of compensating for channel-imposed amplitude and/or phase distortion of the received signal spectrum. The simplest form of equalisation, passive equalisation, may be applied as a compensation filtering, wherever stable propagation frequency distortion is encountered. The one circumstance where passive equalisation would appear to be of use is in tailoring an otherwise inadequate transmit transducer response to improve amplitude uniformity and/or phase linearity. Otherwise, because in the sea, stable propagation would be the exception rather than the rule, some form of active, adaptive equalisation would appear to be inevitable.

In radio receiver design, an early equaliser strategy consisted of combining an amplitude slope equaliser, which compensated for in-band gain slope, and a space diversity receiver which compensates for notches or dips in the received frequency response [10.16]. Such a technique is adequate for use with relatively simple modulation techniques and might thus be commended for underwater applications. More recently, more sophisticated equalisers have been developed. These include adaptive transversal equalisers [10.17, 10.18] and decision feedback equalisers [10.19, 10.20] used either alone or in combination with amplitude slope equalisers and, usually, space diversity receivers.

The equalisation problem is most frequently presented in the context of a main path plus a single interfering path, with the result that the effective channel transfer function H(f) becomes modified as

$$H'(\omega) = (1 + k(t) \exp(-j\omega t)) H(\omega)$$

where k(t) is a time-varying gain-factor for the multipath channel and t is the multipath excess delay. The equalising filter is required to establish a transfer function

$$G(\omega) = (1 + k(t) \exp(-j\omega\tau))^{-1}$$

which may be expanded, according to the Binomial Theorem to yield

$$G(\omega) = 1 - k(t) \exp(-j\omega t) + \frac{k^2(t) \exp(-j2\omega\tau)}{2!} - \frac{k^3(t) \exp(-j3\omega\tau)}{3!} + \ldots$$

This expansion provides us with the basic structure of a finite impulse response filter capable of acting upon a sequence of values sampled at intervals τ from the received waveform. Of course, we may be largely ignorant of the value of τ or, indeed, we may well expect to encounter a more richly multipath environment than that suggested above.

Another way of approaching the equalisation problem is to recognise that the end-result of channel distortion – and the only significant consideration with digital signalling – is intersymbol interference. First, we design the transmitted digit shape (by means of some appropriate pre-transmission filtering) so that, as figure 10.9(a) shows, it exhibits no intersymbol interference. That is, the amplitude of the filtered pulse is zero at time instants corresponding to the adjacent pulse midpoints. This is usually achieved by defining the pre-transmission filter to impose a transmitted pulse spectrum with "cosine rolloff". During transmission, the channel will induce distortion of the pulse spectrum and bring about intersymbol interference, figure 10.9(b). In a digital system, we seek to sample the received bit-stream at the point of some received digit "maximum eye aperture". All adjacent bit-centre values should thus be zero. This can be achieved by means of the transversal equaliser circuit illustrated in figure

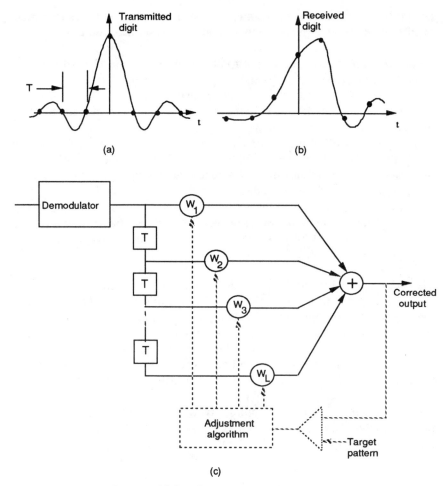

Figure 10.9. The transversal equaliser

10.9(c). For pre-set equalisation, the target pattern is a once-off transmitted training sequence. For adaptive equalisation the target pattern may be obtained directly from the output of the equaliser by passing the output through a slicing or hard-limiting circuit.

10.7 Communication using Parametric Transmission

Quazi and Conrad [10.21] make the suggestion that parametric transmission, because of its ability to establish pencil-beam transmission at relatively low frequencies, with physically small transducers, might have particular advantage in avoiding surface and sea-floor reflections and thus minimise or eliminate the corruptive effects of multipath transmission.

Parametric sonar might be less attractive than Quazi and Conrad suggest, since high directivity at a frequency approaching one of the parametric primaries is, in any case, readily achieved using conventional transmission. One of the remaining advantages of the parametric method is that transmissions using the lower, secondary frequency are less strongly attenuated, in water, than conventional transmissions at the primary frequency. For many applications this advantage would be offset by the poor power efficiency of parametric conversion. Another potential advantage is the possibility of making use of the extreme frequency agility of the secondary frequency. At least in so far as bandwidth is concerned, the absolute width of sweep of the secondary cannot in any case exceed the primary bandwidth. Finally, the added complexity of a parametric projector would increase cost and could adversely affect robustness.

The one potentially significant consequence of parametric transmission could be attached to the aforementioned frequency agility. Thus although the absolute sweep capability is probably very similar, whether a parametric secondary transmission or ordinary transmission at the primary frequency, the proportional sweep capability of the parametric secondary is much larger than that of the innately narrowband primary. Two possibilities then ensue: the application of parametric transmission in a true spread spectrum context (since any other really broadband transmit transducer is difficult to envisage) and also use in a frequency hopping diversity receiver structure, where a proportionately large frequency hop is more likely to result in disassociation from a previously deleterious multipath combination.

References

[10.1] R. Coates and P. Willison, Underwater Acoustic Communications: A Bibliography and Review, *Proc. Inst. Acoustics,* Vol. 9, Pt 4, December 1987, pp. 54-62

[10.2] R. Coates, Acoustic Data Telemetry from Beneath the Ocean Floor, *Proc. IEEE Oceans '87 Conf., Nova Scotia*

[10.3] S.D. Morgera, Digital Filtering and Prediction for Communication Systems Time Synchronisation, *IEEE J. Oceanic Eng.,* Vol. OE-7, No. 3, July 1982, pp. 110-19

[10.4] D.C. Brock, S.C. Bateman and B. Woodward, Underwater Acoustic Transmission of Low-Rate Digital Data, *Ultrasonics,* Vol. 24, No. 4, July 1986, pp. 183-8

[10.5] J. Capotvic, A.B. Baggeroer, K. Von der Heydt and D. Koelsch, Design and Performance of a Digital Acoustic Telemetry System for the Short Range Underwater Channel, *IEEE J. Oceanic Eng.,* Vol. OE-9, No. 4, Oct. 1984, pp. 242-52

[10.6] J.V. Chase, A Tracking and Telemetry System for Severe Multipath Acoustic Channels, *Proc. IEEE Oceans '81 Conf., Boston, Mass.*, pp. 35-39

[10.7] D. Garrood, Applications of the MFSK Acoustic Communications System, *Proc. IEEE Oceans '81 Conf., Boston, Mass.*, pp. 67-71

[10.8] P.O. Kearney and C.A. Laufer, Sonarlink – A Deep Ocean, High Rate, Adaptive Telemetry System, *Proc. IEEE Oceans '84 Conf., Washington, D.C.*, Sept. 1984, pp. 49-53

[10.9] B. Leduc and A. Glavieux, *Long Range Underwater Acoustic Image Transmitting System*, Institut Français de Recherche pour l'Exploitation de la Mer, BP 337, 29273 Brest Cedex.

[10.10] G.R. Mackelburg, S.J. Watson and A. Gordon, Benthic 4800 Bits/s Acoustic Telemetry, *Proc. Oceans '81 Conf., Boston, Mass.*, p. 72

[10.11] C.S. Miller and C.E. Bohman, An Experiment in High-Rate Underwater Telemetry, *IERE Conf. on Eng. in the Ocean Environment*, 1972

[10.12] R.B. Mitson, T.J. Storeton-West and M.G. Walker, Fish Heart-rate Telemetry in the Open Sea Using Sector Scanning Sonar, *Biotelem. Patient Monitoring*, Vol. 5, No. 3, 1978, pp. 149-53

[10.13] R.M. Dunbar, S.J. Roberts and S.C. Wells, Communications, Bandwidth Reduction and System Studies for a Tetherless Unmanned Submersible, *Proc. IEEE Oceans '81 Conf., Boston, Mass.*, pp. 127-131

[10.14] S.O. Rice, Mathematical Analysis of Random Noise, *Bell Sys. Tech. J.*, Vol. 24, No. 46, 1945, Art. 3.10

[10.15] W.C. Jakes, Jr., *Microwave Mobile Communications*, Wiley, New York, 1974

[10.16] Y.Y. Wang, Simulation and Measured Performance of a Space Diversity Combiner for 6 GHz Digital Radio, *IEEE Trans. Commun.*, Vol. COM-27, Dec. 1979, pp. 1896-1907

[10.17] M. Shafi and D. Moore, Further Results on Adaptive Equaliser Improvements for 16-QAM Digital Radio, *IEEE Trans. Commun.*, Vol. COM-34, Jan. 1986, pp. 59-66

[10.18] F. de Jager and M. Christiaens, A Fast Automatic Equaliser for Data Links, *Philips Tech. Rev.*, Vol. 37, No. 1, 1977, pp.10-24

[10.19] P.P. Taylor and M. Shafi, Decision Feedback Equalisation for Multipath Induced Interference in Digital Microwave LOS Links, *IEEE Trans. Commun.*, Vol. COM-32, March 1984, pp. 267-279

[10.20] C.A. Belfiore and J.H. Park, Decision Feedback Equalisation, *Proc. IEEE*, Vol. 67, No. 8, August 1979, pp. 1143-1156

[10.21] A.H. Quazi and W.L. Conrad, Underwater Acoustic Communications, *IEEE Commun. Magazine*, Vol. 20, No. 2, March 1982, pp. 24-30

Index

absorption 20
abyssal plain 52
acoustic centre 16
acoustic communication 171–85
acoustic impedance 9
acoustic intensity 9, 16
acoustic navigation 166
acoustic positioning 166
acoustic telemetry 162, 171–85
admittance, transducer 130–2
ambient noise 28, 90–103
angular distribution function 34
angular distribution of noise 95–100
ARMA analysis 42
array, transducer 121, 136–52, 178–80
array polar response 140–5
array shading 144
array steering 146, 180
attenuation 18, 19
    anomalous 21
    sea-water 19
    sediment 21
auto-correlation 33, 45

backscatter
    surface 109
    volume 107–9
bandwidth
    signal 7
    transducer 125
bathythermograph 5
beamformer 146
beamsteering 146, 180
beamwidth 113, 136, 144, 147
Beaufort Scale 92
Beckmann–Spizzichino loss 27
Bessel equation 86
bistatic sonar 30
blade-rate tonal 101
broadside array 141, 146
bulk modulus 3

caustic 62
cavitation 100, 148
cavitation noise 100
cepstrum 34, 47
circular navigation 167
Collias equation 4
communication, acoustic 171–85
complex cepstrum 50
conductivity 3

continental shelf 52
continental slope 52
correlation function 33, 42, 45
cross-correlation 33, 47
CTD measurement 5
Curie temperature 116
cutoff frequency, mode 78
cylindrical spreading 18

decibel 8
deep-ocean transmission 58–64
depth dependence of noise 94
dipole receiver 140
dipole source 96
directivity
    receiver 113, 147
    source 17
dispersion, waveguide 82, 87
diversity
    frequency 180
    space 178
doppler measurements 168

echo sounder 154
eigenfunction 86
eigenvalue 86
elasticity 3
endfire array 141, 146, 149
energy spectral density 33, 45
energy time–frequency plot 33, 45
equalisation 181

fading behaviour 174
"fast" bottom 14
Fast Fourier Transform 36
ferro-electricity 116
FFT 36
filter-bank spectrum analysis 36
finite energy process 33
finite power process 33
fish tags 162
fishing sonars 160
flextensional transducer 115
free-field spreading 18
frequency
    spatial 7
    temporal 7
frequency hopping 180

Gloria 165
Goll-type transducers 127–9
group velocity 78

Index

half-beamwith   113, 136, 144, 147
Helmholtz equation   85
heterodyne spectrum analysis   34
homomorphic deconvolution   34, 49
hydrophone design   132–4
hydrophone line array   137–43
hyperbolic navigation   168

image interference   70–2, 173–7
impedance, acoustic   9
intensity   9
interference, image   70–2, 173–7
inverse-square loss   18
isothermal propagation   59
isovelocity propagation   55

Janus velocity measurement   169

Langevin projector   117–22
lead zirconate titanate   116
line array   137–43
Lloyd mirror   70
loss
    Beckmann–Spizzichino   27
    NUC   25
    Rayleigh   25
    sea-floor reflection   25
    sea-surface reflection   26
    spreading   18
    transmission   18

magnetostriction   114
mainpath multipath   105
matching, transducer   127
meta-cepstrum   49
mode cutoff frequency   78
mode theory   53
moisture content   21
molecular resonance effects   20
monopole receiver   139
monostatic sonar   30
multipath, mainpath   105
multipath propagation   105, 172–84
multiple matching layers   127–29
multiple source images   66

navigation, acoustic   166
new reference unit   8
noise   28, 90–103
    ambient   28, 90–103
    self   28
    shipping   92, 100–3
    surface agitation   92
    thermal   92
noise level   29, 93
noise spectrum level   92
noise variability, angular distribution   95–100
noise variability with depth   94
noise variability with time   93
non-stationary process   33
normal modes   73–89
NUC loss model   25

parametric communication   183
parametric source   147
particle velocity   1
pattern synthesis, array   143–5
period
    spatial   7
    temporal   7
piezo-electricity   114, 116
Poisson's ratio   119
polar reponse   113, 129, 140–5
polar response measurement   129
poly-vinylidene fluoride   116
porosity   21
positioning, acoustic   166
power spectral density   33, 45
propagation equation   6, 83
propeller noise   100–1
PVF   116
PZT   116

Q-factor, transducer   125
quartz   116

ray coefficient   54
ray-tracing   53–72
Rayleigh distribution   173
Rayleigh parameter   27
Rayleigh reflection loss   25
receiver directivity   113, 147
reflection coefficient
    intensity   12, 24
    pressure   11
refraction   54
resonant transducer   123–7
reverberation   29, 104–10
    surface   30, 104
    volume   30, 108
Rician distribution   173
rigid bottom   77

salinity   3
scattering

surface 109
volume 107–9
sector-scanning sonar 157
seismic surveying 165
self-noise 28
shading 144
shadow zone 62
shelf-sea transmission 65–70
ship noise 100–3
side-scan sonar 163
signal excess 30
sing-around velocimeter 6
singing 101
"slow" bottom 15
Snell's law 10, 53
sonar
    bistatic 30
    mono-static 30
sonar equations 16
sound speed 1
sound velocimeter 3, 6
source directivity 17
source intensity 16
spatial correlation 50
spatial period 7
specific heat 3
spectral density 33
spectral estimation 41
spectrum, wavenumber 50
spectrum analysis
    FFT 36
    filter-bank 36
    heterodyne 34
    Prony 40
spectrum level, noise 92
specular reflection 10
spherical spreading 18
spread spectrum communication 180
spreading loss 18
stationary process 33
Sturm–Liouville problem 86
sub-bottom profiling 159
surface reverberation 30, 104
surface scattering 109
surveying 166–8
synthetic aperture 149

target strength 22
telemetry, acoustic 162, 171–85
temporal period 7
terminal response measurement 130–2
terrain mapping sonar 157, 163
thermocline 60

thin-disc transducer 127–9
time variable gain 158
tonpiltz transducer 122
transducer 112–52
    flextensional 115
    free-flooding ring 122
    Goll-type 127–9
    Langevin-type 117–22
    multiple layer 127–9
    polar response 129
    thin disc type 127–9
    tonpilz 122
transducer arrays 121, 136–52, 178–80
transducer matching 127
transducer terminal admittance 130–2
transducer testing 130–2
transducer tuning 123–7
transducers
    air-gun 115, 159
    at resonance 123–7
    explosive 115, 159
    sparker 115, 159
transformer, design criteria 156
transmission coefficient
    intensity 13
    pressure 11
transmission loss 18
transmission modelling
    shallow-sea 65–70, 73–89
    wedge-sea 70
transmitting response 123–7
transponders 162, 166
travelling wave 6, 83
tuning, transducer 123–7
TVG circuit 158

velocimeter 3, 6
vertical directivity of noise 95–100
viscous friction 19
volume reverberation 30, 108
volume scattering 107–9

wake noise 100
wave equation, 2-D 84
waveguide dispersion 82, 87
waveguide propagation 62, 73–89
wavelength 7
wavenumber 7
wavenumber spectrum 50
window functions 39

XBT 5